HOTEL,
HOSTEL AND HOSPITAL
HOUSEKEEPING

JOAN C. BRANSON
MARGARET LENNOX

HOTEL,
HOSTEL AND HOSPITAL
HOUSEKEEPING

FIFTH EDITION

JOAN C. BRANSON

BSc, MHCIMA

MARGARET LENNOX

Dip Dom Sci

Hodder Arnold

A MEMBER OF THE HODDER HEADLINE GROUP

Orders: Please contact Bookpoint Ltd, 130 Milton Park, Abingdon, Oxon OX14 4SB.
Telephone: (44) 01235 827720. Fax: (44) 01235 400454. Lines are open from 9 am – 5 pm
Monday to Saturday, with a 24-hour message answering service.
You can also order through our website: www.hodderheadline.co.uk

British Library Cataloguing in Publication Data

Branson, Joan C. (Joan Cameron), *1917-*
    Hotel, hostel and hospital housekeeping.
    5ᵗʰ ed.
    1. Residential institutions. Household management – Manuals
    I. Title   II. Lennox, Margaret
    647'.9

ISBN-10: 0 7131 7732 2
ISBN-13: 978 0 7131 7732 9

First published in Great Britain 1965
Second edition 1971
Third edition 1976
Fourth edition 1982
Fifth edition 1988
Impression number   20   19   18
Year   2009   2008   2007   2006

Typeset in 10/11pt Paladium by Colset Private Limited, Singapore
Printed in India for Hodder Education, a division of Hodder Headline,
338 Euston Road, London NW1 3BH by Replika Press Pvt. Ltd.

CONTENTS

PREFACE

The planning, provision and service of accommodation in hotels, hostels, hospitals and similar establishments is a task involving many thousands of people and many thousands of pounds in its operation.

The management of housekeeping departments has advanced rapidly in recent years and requires not only a knowledge of technical skills but also an understanding of the 'tools' of management. A large part of the executive housekeeper's time is taken up with personnel management and the aim should be to have an efficiently run department, with operating costs as low as possible, thus providing a clean, comfortable and safe environment. A well-organized department should contribute significantly to the profitability of the establishment. The housekeeping department should not be considered in isolation as it is an important source of information for many other departments and good co-operation leads to a more congenial atmosphere throughout the establishment.

As in earlier editions of this book the various different aspects of housekeeping are fully explained in order to meet the needs of students and those already working in this field. The content and structure of the book have been influenced by the requirements of the various examining bodies in the hotel and catering field and it is particularly suitable for City and Guilds 708, BTEC and HCIMA courses.

It is not expected that the book will be read at one sitting, therefore some degree of repetition has been introduced where it is considered necessary, and cross-references have been provided in order to ensure a comprehensive understanding of each chapter.

ACKNOWLEDGEMENTS

Without help and encouragement this book would not have been written and we are fortunate in that many hotels, hospitals, hostels and university halls of residence, far too numerous to mention all by name, have opened their doors to us in our quest for information.

We are greatly indebted to the staff of the Hotel and Catering departments of Ealing College of Higher Education and the University of Surrey for their helpful advice and criticism and we appreciate the help our respective families have afforded us, as well as their unfailing interest.

We would like to thank the following for their help in the preparation of the fifth edition: Unilever; Carpet International Ltd; Cimex; and the domestic services management of the Department of Health.

The publishers would like to thank the following for permission to reproduce copyright photographs:

British Red Cross, pp 68 and 69; Royal Garden Hotel and Noeline Kelly, pp 124, 127 and 272, and Royal Garden Hotel, p 269; Heathrow Hotel and Noeline Kelly, p 135; Royal Kensington Hotel and Noeline Kelly, pp 166, 242 and 255; Inn on the Park and Noeline Kelly, p 185, and Inn on the Park, pp 269 (top left, bottom right) and 290; Rufflette Ltd, p 219; Noeline Kelly, pp 231 and 283; Antiference Curtains, pp 222 and 223; Selfridges Ltd and Noeline Kelly, pp 227, 231 and 232; Buckingham Beds, p 232; Tower Hotel and Noeline Kelly, p 239, and Tower Hotel, p 271 (top); Batheaston Chairmakers Ltd, p 249; Heals Contracts Ltd, p 251 (right); Primo Furniture Ltd, Selfridges Ltd and Noeline Kelly, p 251 (left), and Primo Furniture Ltd, pp 248 e) and 257; Victoria and Albert Museum, p 248 a) and b), MAS Contracts Ltd, p 248 c) and Julie Stevens, p 248 d); Hotel Riviera, Sidmouth, Devon, p 253; Parker Knoll, p 254; Concord Lighting Company, pp 269 (top right) and 271 (bottom); Architectural Review and Brecht-Einzig Ltd, p 270; Flowers and Plants Council, p 276; Austin Suite Ltd and Noeline Kelly, p 284; Formica Ltd, p 288; Doulton Sanitaryware, p 290 (bottom); Barnaby's Picture Library, p 293; Garnett College and Noeline Kelly, pp 294 and 295.

A

HOUSEKEEPING

1

THE HOUSEKEEPER AND THE ORGANIZATION OF THE DEPARTMENT

In any residential establishment, be it hotel, hostel or hospital, the basic requirements of the *guest* are for food, drink and accommodation; accommodation being the space and facilities needed for sleeping and/or living.

On arrival the *guest* enters the foyer or entrance hall and gains an impression of the establishment from what can be seen, mainly from the appearance that it presents. From the reception desk the *guest* goes to the lift, staircase or corridor to reach the allocated bedroom, possibly passing the lounge and other public areas. On reaching the room the *guest* probably has time to explore the surroundings more closely; he or she takes a closer look at the décor, the furnishings and furniture, especially the bed, and the cleanliness and comfort of the surroundings. By this time the *guest* is better able to judge the standards of the establishment and to decide whether it is likely to meet with requirements and to provide satisfaction. This first impression is probably obtained before the *guest* has had any food or drink; in fact, the *guest* may neither eat nor drink in the establishment at all, therefore the basic service provided should be a good night's rest in clean, comfortable and safe surroundings. In a hotel the letting of accommodation earns most money and the satisfaction of the guest is of prime importance.

In any establishment there are three departments particularly concerned with accommodation:

1 The reception department, whose staff sell and allocate the accommodation;
2 The housekeeping department, whose staff plan, provide and service the accommodation;
3 The maintenance department, whose staff provide adequate hot and cold water, sanitation, heating, lighting and ventilation as well as maintaining and repairing individual articles and areas within the accommodation operation.

Housekeeping may be defined as the provision of a clean, comfortable and safe environment. It is not confined to the housekeeping department as every member of staff in the establishment should be concerned with the provision of these facilities in their own department, eg the chef 'housekeeps' in the kitchen, the restaurant manager or head waiter 'housekeeps' in the restaurant, and the general manager has overall responsibility.

Housekeeping, domestic administration or accommodation services is therefore essential in all types of establishments, whether hotels, clubs, hospitals or hostels etc, in order that there shall be comfort, cleanliness and service, and all these should be the concern of every member of the establishment.

Job titles

The responsibility for housekeeping may belong to a woman (in the past it was almost entirely confined to women) but men are now becoming more involved, particularly in hospitals, and the exact title of the person responsible varies from one establishment to another.

In hotels, 'housekeeper' is still the usual term but there are small hotels run by husband and wife where the manageress may, in fact, be the housekeeper, although she would obviously prefer to be known as the manageress as this confers a higher status. In others, an assistant manager may organize the housekeeping department.

In hostels and university halls of residence, domestic bursars may be engaged, one of whom is responsible for the housekeeping; but, where the welfare of the residents comes within the scope of this person, the term *warden* is more usual.

In hospitals, 'domestic services manager' is the title used and in boarding schools and homes for children and the elderly, the term 'matron' can be appropriate.

It is one thing to give the correct title to the person responsible for housekeeping in a particular establishment, but it is quite another matter to find a suitable title when referring to establishments in general. There is the choice of one of the following:

accommodation manager,
executive housekeeper,
housekeeper,
domestic services manager,
bursar,
warden,
matron.

However, throughout this book the appropriate titles will be used where possible, and where this is not so, the term *housekeeper* (in italics) will be used, irrespective of the type or size of the establishment or house.

A similar problem arises with regard to the word guest/resident/patient/customer, etc, and so again the appropriate word will be used where possible, otherwise the word *guest* (in italics) will be used generally.

Management in different establishments

The management of the accommodation or housekeeping department will be influenced by such factors as size, type and location of the establishment and no two *housekeepers* will manage their departments in exactly the same way. However, whether the department is large or small, luxury or medium class, for short or long stays, from the commercial or welfare field, management expects the department to be run with the highest degree of efficiency and at the lowest cost.

Whereas the *guest* in any given establishment has the same needs as far as accommodation is concerned, ie a clean, comfortable and safe environment and the need to feel welcome and to be treated cheerfully and courteously by the staff, the same standard of accommodation cannot be expected throughout the wide range of establishments available.

Within the different types of accommodation there are variations in the

size of areas provided
facilities and furnishings provided
services offered to the guest

and standards are generally determined by how much the *guest* is paying.

For the *guest* to be satisfied services must be offered, but for a satisfied *guest* there must be efficiency within the standards decided on and, from the establishment's point of view, the areas and service must be planned so that the accommodation can be put to the best possible use with regard to appearance and earning power.

The standard and tone of the housekeeping department plays a large part in the reputation of the establishment and in determining whether *guests* are satisfied with their stay and, in the case of hotels, wish to return. Whereas the type of service offered differs greatly from one establishment to another, and housekeeping in hotels and expensive clubs may be more specialized than in other establishments, the basic problems of administration are similar. Efficiency in housekeeping should lead to the comfort and well-being of the *guest*: in hotels, clubs and hostels this should lead to a greater or full occupancy; and in hospitals the patient should leave satisfied that, as well as receiving medical and nursing care, his comfort had been considered. Besides this, efficiency in housekeeping should contribute to the saving in costs of labour, cleaning materials and equipment, furnishings and the like, in every type of establishment.

All *housekeepers* should be concerned with the cost efficiency of their departments but an executive housekeeper normally spends a great deal of time on administrative work and she often has to make a real effort to leave the office and get around the department.

The *housekeeper* who has the ability and personality to:

make guests feel welcome
inspire confidence
smooth over difficulties and
train her staff

is an asset in any establishment and should save management many headaches.

The aims of the *housekeeper* are to:

- achieve the maximum efficiency possible in the care and comfort of the *guests* and in the smooth running of the department,
- establish a welcoming atmosphere and a courteous, reliable service from all staff of the department,
- ensure a high standard of cleanliness and general upkeep in all areas for which she is responsible,

- train, control and supervise all staff attached to the department,
- establish a good working relationship with other departments,
- ensure that safety and security regulations are made known to all staff of the department,
- keep the general manager or administrator informed of all matters requiring attention.

There are certain similarities in the areas for which the *housekeeper* is responsible in all establishments. As an example, the areas in an hotel are broadly:

bedrooms, single, twin or double, with or without private bathrooms, suites,
lavatories and public bathrooms,
foyer, lounges, TV and writing rooms,
games rooms and other leisure areas,
corridors and staircases,
cloakrooms,
conference rooms.

(The restaurants, bars and banqueting rooms may, to varying extents, also be the concern of the housekeeper.) There are, in addition, administrative offices, cloakrooms, maids' service rooms, the linen room etc, and the accommodation of living-in staff. (In hospitals there are patient areas.)

A housekeeper's work in any establishment may consist of some or all of the following:

co-operation with other departments;
engagement, dismissal and welfare of staff;
deployment, supervision, control and training of staff;
compilation of duty rosters, holiday lists and wage sheets;
checking the cleanliness of all areas for which she is responsible;
completion and/or checking of room occupancy lists;
dealing with *guests'* complaints and requests;
reporting and checking of all maintenance work;
control and supervision of the work of the linen room and possibly an in-house laundry;
dealing with lost property;
control of all keys in the department;
prevention of fire and other accidents in the department;
care of the sick and the provision of first aid for staff and *guests*;
ordering and control of stores, equipment etc, in the department;
being willing to advise on the interior design of the rooms, cleaning and associated contracts, pest control;
keeping inventories and records of equipment, redecoration and any other relevant details of the department;
floral decorations.

In hospitals the work of the Domestic Services Manager may also include:

management of staff residences;
housekeeping and ward orderly services;

refuse control;
management of patients' clothing in long-stay hospitals.

This list of duties is long and, in order to run the department efficiently, the *housekeeper* has to delegate some of the work to her assistants while retaining overall responsibility. In smaller establishments the housekeeper is much more concerned with the day-to-day routine work and, at times, may have no assistants on duty with her.

The status of the *housekeeper* varies considerably, depending on her experience, length of service, strength of character and personality, as well as on the type and size of the establishment. Depending on her status, her accommodation may be a self-contained flat, a suite or a bed-sitting room, or she may be non-resident.

A housekeeper's attributes should include:

an interest in people and tact in handling them,
a pleasant personality and the ability to converse with all types of people,
an ability to hide personal likes and dislikes, and to be conscientious, fair
    and just,
strictness regarding punctuality and the keeping of necessary rules,
loyalty to the establishment and to her staff,
critical powers of observation,
a sense of humour,
an adaptability and willingness to experiment with new ideas, use initiative
    and take responsibility,
a cool head to deal with any emergencies,
the possession of a strong heart and good feet.

If all these attributes were to be incorporated in one person, they would be a paragon, thus a few of these can be missing if a *housekeeper* has a sense of humour and a pleasing personality.

The *housekeeper* comes into contact with many types of people, management, staff, guests and others, some of whom may be of foreign origin, which can present language problems. Communication then becomes difficult, requiring great patience and tact on the part of the *housekeeper* and other members of the staff.

To sum up, it may be said that while a *housekeeper's* life is a busy one, requiring patience, skill and good humour, it is also very varied and satisfying.

When considering a housekeeping department it must be realized that *guests* may stay in an establishment for convenience, for pleasure or from necessity, and that there are different types of establishments, that many variations within one type exist and that, however similar, no two places will be run in exactly the same way.

Classification and organization

To classify very broadly, there are establishments which satisfy commercial needs and others which satisfy social needs (the welfare sector). In the former category are hotels of various kinds, motels, town and country clubs, boarding houses and holiday camps, and in the latter category are university halls of residence, hospitals, hostels and 'homes' of various kinds.

In many of these establishments the *housekeeper* will have one or more assistants working for her. The assistants will supervise those undertaking the actual cleaning and carry out work delegated to them by the *housekeeper*. These assistants, or supervisory staff, may be given different titles in establishments, as may the operatives (see table below).

Different titles for housekeeping staff

Supervisory staff	*Operational staff*

eg floor supervisors
     assistant housekeepers
     floor housekeepers
     domestic supervisors
     assistant bursars
     assistant wardens
	eg room attendants
	     room maids (chambermaids)
	     domestic assistants
	     domestics
	     ward orderlies
	     ward maids
	     maids
	     cleaners
	     houseporters

As the size of the establishment increases, so the *housekeeper* requires more supervisory and operational staff and one of her assistants may be her deputy or first assistant. Unlike other establishments, in the National Health Service assistant domestic services managers do not supervise staff.

In the case of operational staff: in some establishments there may be several categories with individual names indicating the type of work or place of work, eg room maids (room attendants), staff maids, cleaners and house-porters in hotels, ward orderlies and ward maids in hospitals; while in other establishments, the operational staff may be called 'maids' or 'domestics' or 'cleaners' without any reference to their place or type of work.

Men have always undertaken some housekeeping duties in the Navy on board ship and in the other Armed Forces, as well as in other establishments – there are male room attendants, ward orderlies etc. Throughout this book the term 'maid' or 'room maid' is used without discrimination.

Commercial sector

In the commercial field guests are charged according to the type of accommodation and service they are offered. Thus *hotels* and *motels* may be classified as follows:

1 First-class luxury hotels with private bathrooms, suites and lounges, where the décor is luxurious and provision is made for particular personal services to the guests. This type of hotel will inevitably be very expensive and employ many staff, and there are only a few such hotels in the UK.

2 Good hotels having private bathrooms, some suites, lounges and good

décor, very comfortable but giving less personal service and so less expensive.

3 Medium-class hotels where comfort and furnishings are adequate but personal service is cut to a minimum, thus these hotels will be cheaper.

4 Small hotels with less than 50 bedrooms, where the furnishings and tariff vary tremendously. In such hotels the owners may work as the manager and assistant manager, with a general assistant and few other staff. Some or all staff may combine jobs, eg the room maid may be a relief waitress and the houseporter may serve early morning tea. This type of hotel is not easy to deal with as far as teaching is concerned, but it is by far the largest group in this country and there is great variation in the type of accommodation and service offered.

5 Motels, Post Houses and motor hotels are specialized establishments catering for motorists, situated on main trunk roads. Fewer staff are employed and more 'do-it-yourself' equipment is found. Motels usually have parking facilities close by the accommodation.

Expensive clubs in town or country, providing facilities for recreation and relaxation with some sleeping accommodation, are run very much on the lines of a first-class hotel.

Holiday camps generally consist of chalet-type accommodation and the amount of service varies. In some places cooking facilities are provided for the guests.

Boarding houses are small hotels, generally with simple furnishings and providing little service.

In an hotel it is generally accepted that a **head housekeeper** is one who supervises three or more assistant housekeepers. The housekeeper in any hotel may be responsible for the following members of staff:

Assistant housekeepers (floor housekeepers or floor supervisors) who supervise the maids and carry out work delegated by the housekeeper. While the total number will vary according to the type of hotel, a general rule is one assistant housekeeper for 50 rooms.

Room maids who are responsible for the servicing of the guests' bedrooms, private sitting rooms and often private bathrooms, and who are on call for service to guests. A maid may be expected to service 10–15 rooms in an eight-hour shift eg 7 am to 3 pm or 8 am to 4 pm.

Staffmaids who clean the rooms of the living-in staff.

Cleaners who are usually part-time and whose job it is to clean offices, public rooms, bathrooms and ladies' cloakrooms. In some hotels this work used to be done by full-time housemaids or corridor maids. There are firms which undertake contract cleaning and some hotels use this service, but the housekeeper still 'vets' the work.

Linen keeper who supervises the work of the linen room and who may have several linen maids to assist her in providing clean, presentable linen throughout the house.

Cloakroom attendant who looks after the ladies' powder room.

Houseporters whose work consists of the removal of rubbish, the shifting of furniture, heavy vacuum cleaning and other odd jobs.

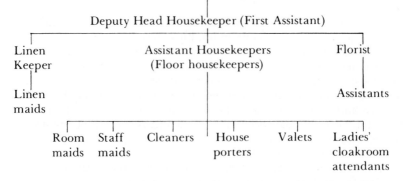

Figure 1.1   Organization chart of a typical housekeeping department in a large hotel

Valets, who usually only work in first-class hotels, are responsible for the valeting of the clothes of the guests and may combine this with some of the less dirty jobs of the houseporter. Valets may be members of the uniformed staff.

A *florist* may be on the housekeeping staff, but in some hotels the house-keeper or her assistants may arrange the flowers and in others there may be contract arrangements.

Window-cleaning is most usually done on contract but large hotels may have *window cleaners* on their staff.

A *general assistant* is one who may be expected to work in any department of a small hotel at any job and so at times may work in the housekeeping department.

For further details of the work of the staff see page 28.

In most hotels it is usual for the manager to confer with the heads of departments regarding matters concerning their departments. The manager's understanding of the work of each department is helpful for all concerned. He (or she) should inform the housekeeper of alterations or arrangements which may affect the running of the department, and she should inform him of any disturbances or unusual occurrences created by guests in their rooms, and any other matters which may require the general manager's attention or advice.

Figure 1.2   Relationship of the housekeeping department to the managerial and other departments in a hotel

Welfare sector

In establishments satisfying a social need a reasonable standard of cleanliness and comfort is expected at the lowest possible cost, and there are tremendous differences in staffing and in services offered throughout the wide range of establishments in this group. Although classification is difficult there are two main areas:

hostels, homes and university or college halls of residence
hospitals

Hostels

Hostels for young people and university halls of residence are medium- to long-stay establishments. Staff are kept to a minimum and students may be expected to make their own beds, keep their rooms tidy etc. Rooms may be cleaned weekly and, apart from in public areas, there may be little or no cleaning at week ends. Some hostels are self-catering, so cooking (and, frequently, laundering) facilities are provided.

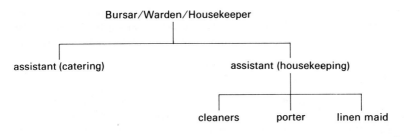

Figure 1.3 Organization chart of a hostel or 'home'

A *bursar/warden/housekeeper/halls' manager* is normally responsible for the general administration of the hostel, the catering and housekeeping, and the maintenance of the building and grounds. Where the establishment is large enough, the day-to-day running of the catering and housekeeping departments is delegated to assistants trained in the particular field. There may also be a warden who will normally combine an academic post with that of being responsible for the welfare and discipline of the residents.

The bursar is responsible for:

Assistant bursars, one or more of whom will supervise the maids and undertake the day-to-day running of the housekeeping department.

Maids/cleaners/domestic assistants who are often part-time and do the work assigned to them by the assistant bursars.

Porters/male domestics, who do the heavier and dirtier work of the house and any other odd jobs.

Linen maid who looks after the work of the linen room.

Window cleaning is normally done on contract, but it may be done by the porters.

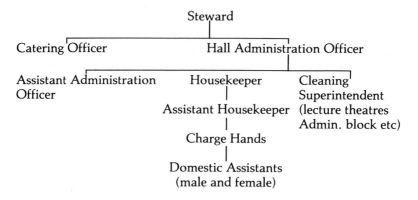

Figure 1.4   Organization chart of a university

On a university campus there may be a number of halls of residence which, from the administrative point of view, are dealt with as one. Terminology of staff varies from one university to another and Fig. 1.4 above is a specific example of one university.

Some *residential homes*, particularly for the elderly, may be run on the lines of a small hotel and there are places where the person in charge, eg the matron, may have nursing rather than housekeeping or catering experience.

Hospitals

In hospitals, the staff residences may be likened to hostels or university halls of residence. The residents' rooms normally receive a weekly clean and sometimes residents clean their own rooms with equipment provided by the hospital.

Hospitals also contain administrative areas, laboratories, training schools, laundries, kitchens and patient areas. Some of these are prestige or high-risk areas, eg operating theatres and renal, transplant and premature baby units. In these and other patient areas great emphasis must be laid on the control of infection.

The extent and scope of the services organized into the domestic services departments in hospitals varies widely throughout the National Health Service from a basic cleaning service to a fully developed housekeeping service. The staffing structure of the domestic services department is different from other establishments and the names of certain grades may be found confusing, eg senior housekeepers in hospitals are not the highest grade and the assistant domestic services manager is at managerial and not supervisory level.

Managerial posts are graded according to a points system which relates floor area to points in the ratio of one point to 500 sq. m. The structure is as follows:

District Advisor/Domestic Services Managers (DA/DSM) are accountable to an administrator for the
– provision of professional advice on a district-wide basis

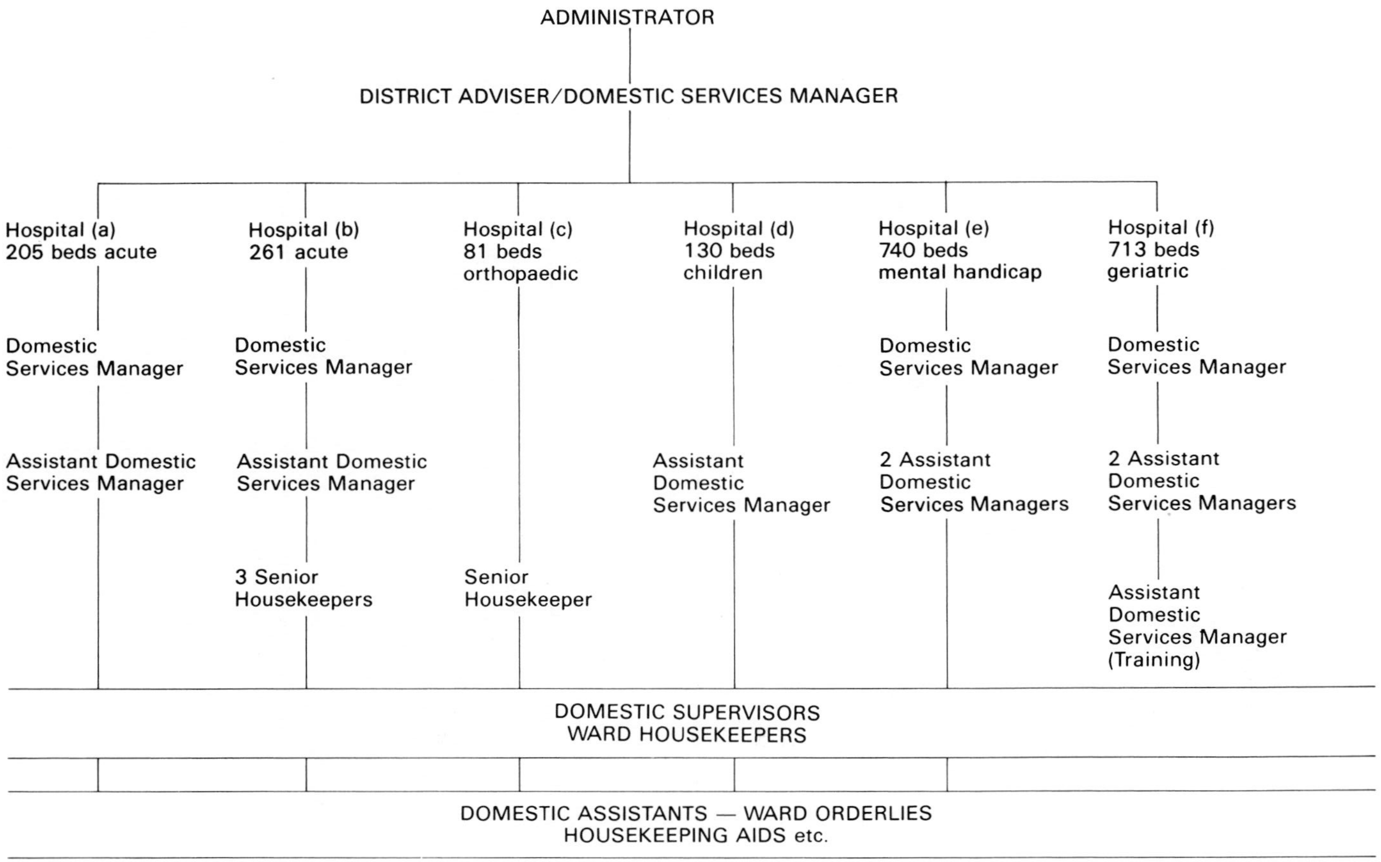

ADMINISTRATOR

DISTRICT ADVISER/DOMESTIC SERVICES MANAGER

Hospital (a)
205 beds acute

Domestic Services Manager

Assistant Domestic Services Manager

Hospital (b)
261 acute

Domestic Services Manager

Assistant Domestic Services Manager

3 Senior Housekeepers

Hospital (c)
81 beds orthopaedic

Senior Housekeeper

Hospital (d)
130 beds children

Assistant Domestic Services Manager

Hospital (e)
740 beds mental handicap

Domestic Services Manager

2 Assistant Domestic Services Managers

Hospital (f)
713 beds geriatric

Domestic Services Manager

2 Assistant Domestic Services Managers

Assistant Domestic Services Manager (Training)

DOMESTIC SUPERVISORS
WARD HOUSEKEEPERS

DOMESTIC ASSISTANTS — WARD ORDERLIES
HOUSEKEEPING AIDS etc.

Figure 1.5 Typical organization chart of the domestic services of the hospital division of a Health District

– management of the domestic services in one or more locations. (A location consists of at least one hospital or other NHS establishment.)

Domestic Services Managers (DSM) are accountable to an administrator and the duties include the organization of the domestic services in a particular location, eg

– drawing up work schedules for the domestic staff, in consultation with nursing staff where appropriate

– recruiting adequate numbers of staff of the grades required within the budget allocated

– maintenance of an adequate standard of cleanliness throughout the location

– control of issue of cleaning materials and equipment

– upkeep of official records on staffing, cleaning materials, training and such other matters as may be required for wages, estimates, procedures etc

– arrangement of staff duty rotas, leave etc

– co-ordination and oversight of training of domestic staff within the location, in conjunction with the DA/DMS and the training and personnel department.

Assistant Domestic Services Managers (ADSM) do not supervise staff in the NHS but are appointed to assist the DA/DSM in the management of the domestic services at a location *or* are accountable:

either to a DSM for assistance in the management of specified domestic services within the DSM's location;

or directly to a general administrator for domestic services at a location too small to qualify for a DSM but of not less than 10 points (5000 sq.m.).

Senior Housekeepers may: (a) be responsible for the domestic services at a level above the supervisory ancillary staff, at a location too small to qualify for a DSM or an ADSM but of not less than three points (1500 sq.m.); or (b) control at junior management level the domestic services in the ward areas, in particular in the organization of domestic staff relieving nurses of non-nursing duties. The areas should be at least six wards and 180 beds, eg six wards of 30 beds or nine wards of 20 beds.

Domestic Supervisors are responsible for the direct supervision of the domestic work in the specific area allotted to them. Generally they would supervise 10 to 20 domestic staff.

Domestic Assistants and *central cleaning teams*, eg floor cleaning, undertake the basic cleaning tasks throughout the hospital.

Ward Orderlies perform those cleaning tasks in the ward areas which are not undertaken by the cleaning teams, as well as some non-nursing tasks.

In the National Health Service there is, at present, a system of competitive tendering whereby the domestic and other support services are put out to tender. The intention is to get better value for money out of available resources. The in-house offer competes with offers from contract cleaning firms. The contract runs for about three years, but can be extended to five years.

Detailed specifications of the work required (see page 51) are drawn up by the in-house domestic services manager, work study experts or a junior administrator, after consultation with senior nurses and heads of departments. The in-house tender is prepared by the district advisor with the help of the treasurer and this is considered with the other tenders put in. On acceptance of an offer the tendering figure becomes the working budget.

A contract cleaning firm's internal staffing structure is similar to the pattern of in-house structure already described, eg manager, assistant managers, supervisors etc, and the contractor may decide to pay at least those rates of pay laid down by the Whitley Council.

Control of the quality of work carried out by the contractor's staff is the responsibility of the contractor through his management structure.

The role of the Health Authority is to ensure that the contractor's quality control procedures provide the service required and to do this a monitoring officer is appointed. In-house services are similarly subject to monitoring.

The monitoring officer undertakes routine and random inspections throughout the establishments and check-lists are filled in. Complaints from heads of departments (wards etc) should be investigated and the necessary action taken. Full details should be recorded and copies sent to the contract manager. Any unsatisfactory performance, as well as the action taken by both the monitoring officer and the contract manager, should be recorded on an appropriate form. Both the monitoring officer and the contract manager should sign the form which should be sent to the authorized officer, who is usually an administrator.

Computerized monitoring systems are now used in some hospitals.

Co-operation

The housekeeping department is just one department in any establishment working towards the satisfaction of the *guest*, and each department is dependent on others for information and/or services if its work is to be done efficiently. During the course of work the *housekeeper* comes into contact with practically every other department and if her work and that of her colleagues is to be unhindered, friction between one department and another kept to a minimum, and the establishment to run as smoothly as possible, there must be close inter-departmental co-operation. Depending on the type and size of the house, the work in each of the other departments may be small enough to be dealt with by an individual or so large that there is a head of department, but in all cases there must be good liaison and communications are helped by a bleep system and computer visual display units (see page 88).

Owing to the many different types of establishments it is not easy to give details of co-operation which would apply in all cases. For this reason, a large hotel has been chosen as an example and the following are the departments with which there should be close co-operation.

Reception

Reception and housekeeping are both concerned with rooms – the former with letting and the latter with preparation and later servicing of the rooms.

For this to be done efficiently there must be a constant exchange of information between the two departments and each must understand the work and possible difficulties of the other.

The housekeeper relies on the receptionist to let her know on which days guests are arriving or leaving, when VIPs are expected, moves required and when special requests have been made for cots, bed boards or baby sitters etc, in order that guests' special requirements may be anticipated and complaints avoided. It is helpful in the organization of the housekeeping department if she is told of group arrivals and late departures.

It is usual for hotels to state that rooms must be vacated by noon on the day of departure but, owing to the increased number of guests arriving early for one reason or another, new guests may arrive before the rooms have been serviced. To avoid guests being shown into untidy rooms the housekeeper should notify the receptionist of 'ready rooms' as soon as they become available and when moves have been completed. She also notifies them when rooms are 'taken off' for redecoration, and again when they are 'put on'. At certain times of the day the housekeeper will let the receptionist have a control sheet (occupancy list, housekeeper's report or vacant room list) so that the receptionist may check the accuracy of the room booking board or chart. The housekeeper often knows of charges which need to be put on a guest's bill, eg early morning teas, the hire of hair driers etc, and the receptionist should be told of these promptly so that the bill is made up ready to be presented to the guest when required.

With this prompt exchange of information there is less likelihood of friction arising between the two departments and in large modern hotels tape recorders, telewriters, room status boards or computers may be installed to facilitate communications.

Maintenance

In the course of the day the housekeeper finds many items requiring attention, such as dripping taps, WC cisterns not flushing, faulty electrical plugs or broken sash cords. These faults should be reported as early in the day as possible and it is helpful if the same types of repairs are grouped together on the maintenance list, or even the repairs in one area grouped together. The housekeeper should co-operate by getting room doors unlocked promptly when repairs are being done. Agreement has to be reached as to when maintenance staff are available for redecoration and the housekeeper told how long this is likely to take. Maintenance should be given rooms ready stripped when redecoration is to take place and furniture removed for repair should be labelled. With co-operation jobs get done more quickly.

Urgent repairs are normally reported to maintenance by telephone and if a good relationship exists between the two departments it is more likely that urgent repairs will be dealt with promptly instead of being added to wait their turn on the list – which can lead to friction between the housekeeper and the receptionist, as well as with the maintenance staff. However, it should be remembered that maintenance is required in other departments besides housekeeping.

Restaurant

Co-operation here is mainly concerned with the linen. While the linen keeper, under the supervision of the housekeeper, needs to have sufficient stock to meet the demands of the restaurant, the restaurant manager should ensure that the times for exchange of linen are respected, and that linen is not mis-used. The housekeeper should be notified of banquets as soon as possible, not only because of the requirements for linen but because in some hotels the housekeeper is responsible for the flowers as well. Co-operation is particularly necessary where there is a floor waiter service, so that friction does not arise over such trivial matters as waiters not collecting trays from the rooms, waiters leaving trays on the corridors, or causing extra work through careless spills on carpets.

Kitchen

The same co-operation is necessary regarding linen as for the restaurant. In addition, a happy atmosphere between the chef and the housekeeper makes one important aspect of staff welfare, ie food, much less of a problem, as complaints may be discussed on a more friendly basis.

Accounts

Wage packets are made up from the information received from the house-keeper regarding hours worked, holidays taken, days lost due to sickness etc, and where this is accurate and punctual, it is hoped that the staff of the house-keeping department will not be kept waiting unnecessarily in a queue for their wage packets. The housekeeper should see that the income tax forms of new staff, notification of staff leaving and of any accidents, petty cash slips and checked invoices, are handed over promptly.

Hall porter

Co-operation with the head porter is necessary regarding lists for early morning teas and calls, the prompt removal of luggage from vacated rooms, and the willing loan of his staff when houseporters are not available. It is helpful if luggage porters tell the housekeepers when luggage has been taken from or to a room so that they know when a room has been vacated or let.

The housekeeper and/or the linen keeper should co-operate by ensuring that the linen room key or a supply of linen is available for the night porter (when this is the house custom) in case of an emergency during the night. (In small hotels the hall porter may also be in charge of guests' dry cleaning.)

Security

Co-operation here is mainly concerned with the prevention of fire and theft and the safe keeping of keys and lost property. There are so many security hazards on the 'floors' that liaison is particularly important and the house-keeper co-operates by endeavouring to see that her staff are aware of them and by reporting anything of a suspicious nature.

Laundry (in-house)

This applies where the laundry is under the supervision of a laundry manager.

Without clean linen the maids cannot operate. At times when there is full occupancy the housekeeper needs a fast turn-round of linen from the laundry, but she should not always be making emergency demands on them. She should co-operate when possible with their normal scheduling and, in return, the laundry should provide an acceptable service in regard to the laundering of the linen.

2
ROUTINE METHODS OF WORK

The housekeeping department plays such a large part in the economic running of an establishment that a good *housekeeper* should, while maintaining efficiency throughout the department, save as much on operating costs as possible. Methods which may be employed to achieve this include work study, the forecasting of labour requirements in relation to room occupancy, and budgeting; these are dealt with more fully in Chapter 6.

The organization of the housekeeping department is therefore an important job, and lack of good organization may cause repercussions throughout the establishment.

If staff are to work efficiently, and cleaning standards are to be maintained, the housekeeper should devise suitable cleaning methods and frequencies and should provide orders of work, suitable equipment and cleaning materials for the work to be carried out.

Cleaning

Cleaning is the removal of dust, dirt and any foreign matter, such as dead flowers, stains, the contents of ashtrays and waste-paper baskets, etc. It is necessary for hygienic reasons, for the sake of appearance and to prevent deterioration of the articles or surfaces.

Dust

Dust consists of loose particles, some of which may cause abrasion, eg grit. Most dust is airborne and may be moved around by air currents, movement of people, equipment and other articles and will later settle on any surface – horizontal, vertical, hard or soft. It is essential that during the removal of dust the particles are collected and not merely shifted from one place to another.

The removal of dust is an important part of the work of the housekeeping staff, and its collection may be by sweeping, mopping, dusting or suction.

Sweeping is not the most efficient, hygienic or modern way of removing dust as so much becomes airborne, and in many cases sweeping has been replaced by the use of the vacuum cleaner.

If not carefully carried out, *dusting* with a dry duster or a dry mop merely moves the dust from one place to another and does not collect it. The collection of dust may be helped if the duster is used damp and the mop is impregnated with a special dressing.

When using a vacuum or *suction* cleaner dust is collected straight into the container and no dust is airborne, which makes it the most efficient means of removing dust. When several methods of removing dust are required in one area because of the variety of surfaces, vacuum cleaning the floor is the last job done.

Thus a room having a polished surround to a carpet square may be cleaned in either of the following orders depending on the equipment available.

1 Dust all surfaces		1 Brush upholstery and surround
2 Mop surround	*or*	2 Dust all surfaces
3 Vacuum clean upholstery and carpet		3 Vacuum clean carpet

Dust which has been collected may be wrapped in paper and put down a rubbish chute or into a plastic or paper sack or dustbin which will later be emptied by a houseporter; if on a mop or duster, these should be washed when necessary.

Dirt

Dirt is dust or other material which adheres to a surface by moisture or grease and which may normally be removed by washing, mopping, shampooing or scrubbing, when friction must be applied in conjunction with hot water, a detergent and possibly an abrasive. However, if it is very greasy dirt a grease solvent, such as white spirit, may be necessary for its removal. Once the dirt has been loosened it may be picked up by means of a cloth, a mop or a vacuum drying machine (ie wet pick-up machine).

Cleaning routines

Any establishment has to present an inviting, clean and well-cared-for appearance at all times, and the cleaning should be carried out at a time when it will cause as little inconvenience as possible. Thus the public rooms and offices are cleaned by maids before breakfast when there is less activity. In some establishments, the public rooms are cleaned during the night by the night porter and only the final dusting is left for the maids. Contract cleaners may be employed in some instances, working either late at night or first thing in the morning. During the day, in hotels, the lounge is normally looked after by the lounge waiter or one of the uniformed staff regarding papers, ashtrays and cushions etc, and the housekeeper inspects the area at intervals.

A maid works from a service room or pantry, where she keeps her equipment (generally marked with her name, number of the floor or section), cleaning agents and other necessities for her work. She may share the room with one or more other maids and when their work is finished the door should be locked. The room should be easily cleaned, with as few things on the floor as possible.

Depending on the establishment, early morning tea may be served from this room, a store of linen kept (sufficient to re-sheet the section or sections), hot water bottles filled and cloths washed and dried.

There is little work that can be done 'on the floors' before breakfast, other than calls and the serving of early morning teas where this is still done in hotels.

The service of early morning tea

In some hotels early morning teas are served by floor waiters, while in the majority, these days, there are tea-making facilities in the rooms (comfort trays), but room maids may carry out this duty from their service rooms. A tea book is made up by the night porter, and a list is given to the room maid when she comes on duty early in the morning; to this list she adds any extra teas that she may serve in order that the charge may be put on the guests' bills. Each maid has a small stock of tea and sugar which is replenished daily according to her list, but milk is fetched each morning from the still-room or kitchen. In some hotels the night porter may have the water boiling in the maid's service room to enable her to make tea straight away if necessary, and any teas requested before the room maid comes on duty will be dealt with by him.

The trays are often kept laid up in the service room ready for use after the crockery has been washed up. The teaspoons, which can have a habit of disappearing, are carefully guarded by the room maid and either cleaned by her or taken occasionally to the plate room.

For the service of early morning tea, the room maid:

1 Knocks on the door and if no reply, waits a few seconds, knocks again and enters.
2 Switches on the light if it is dark.
3 Says 'good morning' cheerfully.
4 Places tray on bedside table with newspapers if ordered.
5 Makes sure guest is awake.
6 Draws curtains if requested.
7 Closes the door gently.

In some hotels the maids are responsible for floor service throughout the day. In this case it is usual for the prepared tray and food to come direct from the kitchen, the 'dirties' are sent back there and the room maid is not concerned with the washing-up.

Corridors and staircases are not normally cleaned before breakfast in case *guests* might be disturbed, and maids should realize that noise, eg shouting, raucous laughter, the banging of equipment and the clatter of crockery, must be avoided at all times.

Guests should be inconvenienced as little as possible and their belongings should only be moved when necessary. Drawers should not be opened by a maid in an occupied room, although clothes may be hung in the wardrobe. On no account should a maid try on jewellery or make use of any of the *guests'* personal belongings, such as cosmetics. Newspapers, unless in the wastepaper basket, should not be thrown away.

It has been customary for maids to leave the door ajar while they are working in a room (this gives the appearance that everything is above board), and as the door is normally opposite the window, care should be taken to avoid articles being blown off the dressing table by the curtain. However some hotels now consider that for security reasons the maid should have the door closed. Equipment and cleaning agents should not be left untidily in the corridor for people to trip over, nor should they ever be placed on the bed or on upholstered furniture and, where it is used, a trolley outside the door indicates that the maid is in the room. In some hotels, there may be a 'maid finder' device outside each door, and the maid operates this to denote that she is in the room.

Rooms at any one time may be occupied (let), vacated (guest has left) or vacant (not occupied last night) and the amount of cleaning given to each room will vary.

All rooms require a comfortable, clean and presentable appearance and the work to be done in an *occupied room* will be enough to maintain this image. This work is known as a daily clean.

In a *vacated room* all signs of the previous guest have to be removed and the room made ready for a new arrival; this entails a special clean and obviously takes longer than a daily clean.

A *vacant room* is one that has not been occupied since the last clean and its appearance is possibly maintained by daily dusting only.

Daily routines

A daily routine is normally carried out by a maid during one visit to a room; but in some instances she may do one job throughout a number of rooms before returning to the first room to do the next job throughout, and so on until the work is completed. This cleaning routine (block cleaning) is more suited to establishments where it is known that the rooms are likely to be empty for at least all the morning and this method does present security risks.

It is usual to give a *special* or *more thorough clean* to occupied rooms when guests have stayed for some days, as well as to vacated rooms before re-letting. This entails giving attention to carpet edges, upholstery, furniture, paintwork etc. Like the daily clean, this special clean may be carried out by the maid on one visit to the room, as would be required in a vacated room, or it may be more convenient in some cases to add one or two jobs to the daily routine, so completing the special clean within a few days. For example:

Mon. – daily work and polishing furniture,
Tues. – daily work and carpet edges and upholstery,
Wed. – daily work and paintwork.

In hotels the number of occupied, vacated and vacant rooms in any one section varies daily, while in hostels, homes and long-stay hospitals the numbers remain relatively constant. Thus in hotels particularly, work loads vary from day to day.

When considering the timing of room cleaning, some cleaning is necessary every day, more thorough cleaning should take place approximately weekly, and a very special clean is necessary periodically.

Spring cleaning

Periodic cleaning is often referred to as 'spring cleaning' or annual cleaning and is carried out at predetermined frequencies, depending on the policy of the establishment. It may be monthly or annually, or any frequency in between. Obviously the less frequent that the periodic clean is, the more extensive it needs to be. All rooms in time require a complete overhaul and many establishments have a spring or annual cleaning programme which may coincide with the planned schedule for redecoration or the washdown of walls and ceilings.

The annual clean in a seasonal establishment is carried out when it is closed and in others at convenient times depending on occupancy. These times will vary according to the type of establishment, eg resort hotels are probably busier at weekends, city hotels during the week, and the exact dates will have to be agreed with reception, maintenance and contract cleaning; security and the laundry (where it is on the premises) will also have to be informed. In some cases, generally depending on the work to be carried out, a whole floor or wing of rooms may be taken 'off' for several days, so that the cleaning may be undertaken.

In *hostels* and *halls of residence* the quiet period is obviously during the vacations but problems arise when rooms are let for conferences and holiday-makers, so careful planning is necessary.

In *residential homes* there may be no quiet periods when a resident is away or could be moved to another room, so the cleaning may have to be planned over several days – disturbing the resident as little as possible.

In *hospitals*, frequent and thorough cleaning of such items as beds, bedding, lockers etc, is essential to prevent the spread of infection so many of the tasks are ongoing rather than periodic and are often dealt with particularly when a patient vacates the ward. This is called terminal cleaning and spring cleaning in its accepted sense is generally only done when the ward is completely stripped and closed for redecoration or washdown. More extensive cleaning of floors and walls is possible in single rooms and the extent to which this cleaning will be carried out depends on whether it is an isolation or high risk area.

Orders of work

In order to help maids in their work and the *housekeeper* in training them, it is possible for orders of work, incorporating work simplification, to be planned. It is a simple matter to plan an order of work for cleaning a specific article, eg a wash basin, but difficulties arise when it comes to rooms, as they vary so much, in addition to the fact that they may be occupied, vacated or vacant.

In general it should be remembered that:

sweeping with a broom is done before dusting, and dusting before vacuum cleaning;
dusting is done from high to low;
vertical surfaces, eg walls, need sweeping occasionally with a wall broom or suction cleaner;

bending with stiff knees should be avoided;

where there are alternative methods of cleaning, the least harmful ones should be used;

cleaning methods should be efficient but also economical of time, labour and cleaning materials.

In *hospitals* in particular:

damp dusting is used to control dust;

mop sweepers are used (instead of brooms) with heads which can be sterilized or disposed of after each cleaning session where necessary;

spray cleaning of floors (using a high speed machine) can be used to give a highly resistant finish;

dust control mats and non-absorbent surfaces are used where possible;

walking backwards is advisable when wet cleaning or applying polish to a floor;

coded or different coloured pieces of cleaning equipment are used in special areas;

noise levels of equipment should be less than 70 dB;

suction cleaner filters should have an efficiency of 60 per cent (BS 5415).

A very simple *order of work* for any area could be:

1 Open windows where possible.
2 Remove litter and dirty crockery, etc.
3 Attend to main jobs, eg bed, fireplace, etc.
4 Sweep if required.
5 Dust and if necessary dry mop.
6 Vacuum clean carpet and upholstery.
7 Spot clean the carpet and paintwork as necessary.
8 Survey the room.

The following are suggested orders of work for the cleaning of several articles and different types of rooms, but it must be stressed that there will probably be adjustments necessary in differing circumstances. (For bed making etc see Chapter 18 Beds and bedding.) Before starting the work it is expected that maids will have been instructed regarding the necessary 'tools' for the job.

Note Only a few articles are mentioned here and others are covered in their respective chapters.

To clean a telephone

1 Dust daily and wipe ear piece free of grease.
2 Occasionally clean the dial and disinfect ear and mouthpiece.

To deal with a TV set

1 Remove plug from wall to disconnect electricity.
2 Move set as little as possible.
3 Dust all over.
4 Use damp cloth to clean screen.

5 Report frayed flexes and other defects.
6 Leave set unplugged.

The care of an electric blanket

1 Remove plug from wall to disconnect electricity.
2 Keep blanket as flat as possible, not crumpled up or bent unduly.
3 Avoid getting wet.
4 When necessary, send protective covering to the laundry.
5 Report frayed flexes and other defects.
(*Note* Electric blankets should be returned to the manufacturer for cleaning and servicing.)

To clean mirrors and glass surfaces

1 Dust daily.
2 Wipe with damp cloth when necessary.
3 Polish with lint-free cloth.
4 Treat frame according to kind.
(*Note* Hair spray marks can be removed with a cloth moistened with methylated spirit or spray-on furniture polish.)

To clean cork or rubber bathmats

1 Wipe daily with a damp cloth.
2 When necessary, wash and rub using fine scouring powder.
3 Rinse and stand upright to dry.

To clean a wash basin (lavatory basin) or bidet

1 Remove hair, fluff, etc, from waste, chain and overflow.
2 Wash and dry toothglass.
3 Clean basin, pedestal and surrounds with swab and scouring liquid, paying particular attention to soap wells and round the base of the taps.
4 Rub up taps, chain and plughole and dry basin.

To clean a bath

Baths are cleaned in a similar manner to wash basins, including the surround, but there is more likelihood of scum and staining, making cleaning more difficult.

If there is a shower, the fittings and curtain rail should be cleaned; the curtains should be wiped and left hanging inside the bath.

To clean a WC

1 Flush pan, brush well and flush again.
2 If pan is still stained, use toilet cleanser and give time for it to work.
3 Brush and flush again.

4 Wipe pedestal, seat, lid and surrounds with a suitable cloth, and dry.
5 Check for toilet paper and leave a spare.
(Further details regarding sanitary fitments are given on pages 158–60.)
Note Chambers are not now provided in bedrooms unless specifically requested. When they have been used the covered chamber is carried and emptied in a WC, ie toilet.

All or some of the above may be housed in a **bathroom** and there may be other fittings and furniture such as a sanibin, stool, towel rail, etc which should be cleaned according to their kind. In addition there are other surfaces, eg shelves and walls, and these show soiling from condensation and dust, especially talcum powder, and so need to be cleaned with a damp cloth. The floor should be of some easily cleaned material which may be washed daily.

Daily cleaning of a bathroom (with tiled walls and floor)

1 Open window if possible.
2 Remove soiled linen and empty sanibin.
3 Clean WC.
4 Clean bath, shower and wash basin.
5 Wipe remaining fittings and surfaces, including walls, mirrors etc.
6 Put out clean towels and bathmat. Check for soap, toilet paper, lavatory cloth and paper bags for sanitary towels.
7 Wash the floor.

Order of work for the daily cleaning of an occupied room

(Room has a wash basin and a carpet square.)

1 Open window; if necessary remove early morning tea or breakfast tray.
2 Strip bed.
3 Empty ashtrays, waste-paper basket and generally tidy room.
4 Attend to wash basin; fold towels and check for soap.
5 Make bed.
6 Adjust window.
7 Dust all furniture and fittings.
8 Mop surround.
9 Carpet sweep or vacuum clean carpet square.
10 Survey room and close door.

The cleaning of a private sitting-room in an hotel

This is one of the jobs a maid does before breakfast, bearing in mind that she must be very quiet and possibly not use the vacuum cleaner. The work that she will do in this room is removing rubbish, straightening chairs, bunching up cushions, etc and dusting; if necessary, the carpet can be dealt with later in the day.

The contents of the sideboard and/or cocktail cabinet in a guest's room are not usually the concern of the room maid, but of the floor waiter. Many hotels now have mini-bars in bedrooms or private sitting rooms and the checking of the contents for billing purposes and the replenishing is done by special staff.

Order of work for the special cleaning of a vacated room

(Room is close carpeted and has a private bathroom.)

1 Open the window; if necessary remove early morning tea or breakfast tray.
2 Look for lost property and wipe out drawers, inside of wardrobe and check for coathangers.
3 Strip the bed, remove soiled linen including towels.
4 Empty ashtrays, waste-paper basket, etc.
5 Make bed with clean linen.
6 Adjust window.
7 Sweep carpet edges and upholstery if no suitable vacuum cleaner.
8 Remove marks from paintwork and attend to mirrors, furniture and fittings, including all ledges, pictures, lights, telephone, TV and radio, wiping, dusting and polishing as necessary.
9 Refill folder, replacing cards and literature if at all marked, making sure that Bible and telephone directories are conveniently placed.
10 Attend to the bathroom:
   Wash basin, bath, WC (see page 23).
   Wipe or dust all surfaces.
   Put out clean towels, soap and toilet paper.
   Clean floor according to kind.
11 Vacuum clean upholstery and carpet edges if suitable vacuum cleaner.
12 Vacuum clean carpet.
13 Survey rooms and close door.

It is best to give the bed plenty of time to air and so the bathroom could be serviced before making the bed. However, in hotels, it is usual to make the bed early in the procedure in order that the room looks tidy as soon as possible. Ideally, the assistant housekeeper should have checked the vacated room for lost property, maintenance and missing articles before the entry of the maid.

In hospitals, when a patient vacates a ward the particular bed area is thoroughly cleaned, eg bed, bedding, locker, chair etc.

The cleaning of staircases

Stairs may be close carpeted, or the carpet may only cover about two-thirds of the stair, in which case there are two surfaces to clean. By using suitable attachments to a vacuum cleaner the two surfaces and skirting board may be cleaned together.

Uncarpeted stairs should be swept daily and washed and/or scrubbed according to the material, when necessary. If a staircase has to be washed while people are using it then, provided that it is wide enough, half should be done at a time, enabling the people to walk up and down on the dry part of the staircase.

It should be remembered that where the side of any staircase is open, dust and dirt may fall through, therefore when sweeping the dust and dirt should be swept towards the wall on each stair.

All bannisters and handrails should be dusted before vacuum cleaning, or after sweeping, and washed or polished occasionally according to material.

Stair rods of brass or polished wood may still be used but nowadays the stair carpet may be held firmly in position by the use of the 'tackless gripper' (see page 187), which eliminates the use of rods and makes cleaning much easier.

In a hotel cleaning of the lifts is rarely the concern of the housekeeping staff but is usually done by the uniformed staff.

The cleaning of utility rooms or kitchen/dining areas in hostels (amenity areas)

In hostels and halls of residence the cleaning of kitchen/dining areas for the use of residents follows the normal cleaning process, but the area does present problems for the cleaner.

In a given kitchen/dining area there will be a refrigerator and food storage cupboards and these are the responsibility of the users, as is the washing up. The cleaner, therefore, cleans the area including the sinks, draining boards and cookers. However, problems arise when the sink is full of dirty crockery and the cooker has been carelessly used, which means that more than the allotted time is needed for cleaning.

The cleaning of computer rooms

1 It is essential for dust levels, humidity and temperature to remain within specified levels.
2 No dry dusting or sweeping – use suction cleaner, damp impregnated mops and mittens.
3 Avoid use of aerosol polishes with silicones.
4 Wet cleaning with the minimum amount of water by damp mopping or scrubbing. Rinse thoroughly.
5 Detergent should be neutral synthetic and leave no resinous deposit on the floor.
6 Avoid use of soap, polish and seals.
7 It is essential that no film builds up on floor surface.
8 Scrubber/polishers should be suppressed and fitted with a suction unit.
9 Do not use floor polish unless specially agreed and then use only a metallized emulsion polish.
10 Clean items of computer equipment only with the users' agreement.
11 Dust control equipment should be cleaned outside the area.

Cleaning of leisure areas (see page 292)

The swimming pool and fitness rooms may be cleaned by the staff involved in running the centre, as they often have time while attending it. Lounge areas are the responsibility of the housekeeper.

Order of work for the annual, spring or periodic cleaning of a bedroom

1 Ventilate room.
2 Strip bed and deal accordingly with linen and bedding.
3 Strip the room of loose furnishings and small articles, including lamp shades, pictures, etc.
4 Vacuum clean and cover bed and upholstered furniture.
5 Vacuum clean carpet, then cover or take up.
6 Wipe or wash furniture inside and out.
7 Stack and cover furniture or remove from room.
  The room can now be redecorated or washed down.
8 Sweep walls and floor.
9 Wash paintwork and have windows cleaned.
10 Thoroughly clean wash basin.
11 Have carpet relaid or uncovered and vacuum clean it, shampoo if necessary.
12 Have curtains rehung.
13 Remove dustsheets, reline drawers, polish furniture and if necessary reposition it.
14 Return cleaned small articles and put in place.
15 Make bed with clean linen and bedding.
16 If there is a surround, mop and polish it.
17 Finally dust, mop if necessary and vacuum clean carpet.
18 Survey room and close door.

During the preliminary preparation, repairs of all kinds (whether to furniture, floors, plumbing or electrical fittings) will be noted and either dealt with *in situ* or the articles removed to be repaired elsewhere.

In an hotel, houseporters or valet-porters will help the maids with some of the jobs done during spring cleaning; in other establishments there are generally porters or handymen available to help with high and heavy jobs.

Annual cleaning of lounges is carried out in a similar way.

3

THE STAFFING OF THE DEPARTMENT

The hours that a housekeeping department is manned will vary considerably and will be determined by the areas to be maintained by the department and the type of service to be given. Not only will these factors influence the organization of the department as far as coverage is concerned but they will, of course, have a bearing on the number of staff and the amount of work that can be undertaken by each employee (see p. 75).

The type of service offered to the *guest* is a question of house policy and/or economics, and the housekeeping department must be geared to meet the demands made on it.

In hotels and similar establishments, the guests pay for service and the hours the housekeeping department is manned must cover not only the time taken for the actual cleaning of the areas but must enable the service to be given to the guests as and when it becomes necessary. The hours of coverage may therefore appear lengthy, but it is necessary that someone is available at the most likely times to meet the guests' requirements.

In an hotel an average coverage may be from 7 am to 10 pm. However, there are hotels where the staff start and finish at other times, earlier or later, and even give a 24-hour service; while in the smaller hotels there may be times at which there are no housekeeping staff on duty and any queries will be dealt with by the general assistant, receptionist or manager.

In university halls of residence, nurses' homes, etc, the residents do not receive the same amount of personal service as guests in hotels and the house-keeping staff are more concerned with keeping a reasonable standard of cleanliness at the lowest possible cost; these places are therefore staffed for cleaning and once this is completed it is not so necessary for the housekeeping department to be manned.

It is possible of course that in any establishment there is work which can only be carried out by day and other work which is more conveniently done in the evening. Therefore there will be tremendous variations in the hours covered by the housekeeping staff, not only from one type of establishment to another but within a group of similar establishments.

Staff duties in hotels

The busiest times in the housekeeping department are in the mornings, when resheeting and essential cleaning of the rooms takes place. This work cannot normally be commenced before about 8.30 or 9 am but some of the room maids may have started their duty before this to serve the early morning teas. There are hotels where early morning teas are not served by the housekeeping staff but by floor waiters, or there may be tea making facilities in the rooms, and in such places maids may come in at 8 to 8.30 am rather than the earlier hour of 7 am.

Where maids 'turn down', draw corridor curtains and are on call to attend to guests' requirements, evenings may also be a busy time but there is an increasing number of hotels where no 'turning down' is done in the evening. In these hotels there will be a minimum of maids on duty to deal with unexpected and late departures and, in the case of Post Houses, motor hotels and similar places, the work produced by rooms being let more than once in 24 hours. There are also hotels where no maids are on duty in the evening and requests and complaints from guests will be dealt with by the housekeeper (if on duty) or by the receptionist.

The afternoons are for completing the morning's work and carrying out extra jobs allocated by the housekeeper.

In some hotels guests spend a lot of time in their rooms, having breakfast in bed, resting in the afternoon, changing at a leisurely pace in the evening, and requiring the odd needle and cotton, flower vases, the use of the iron and board etc, and this impedes the work of the room maid; so in these hotels maids will be given a smaller section to service, and when they are not actually working they will be in their service rooms on call.

There are other hotels geared to business people, where the rooms may be of the studio type and the guests may have business associates or friends in their rooms at any period of the day; this may mean that the guests will ask for the room to be serviced by a particular time, and if there are several such requests this may present difficulties for the maid.

Package tours may disrupt the work because of their arrival or departure at awkward hours and rooms may have to be serviced quickly for new arrivals.

In still other hotels, and particularly transit hotels, there will be many guests staying only one night, who arrive late and leave early, and so require practically no personal service; there may be other guests who stay longer but spend little time in their rooms, and thus in these hotels the room maid's work will be very straightforward and without many interruptions.

Other problems which the maid may encounter in any hotel during the course of her work include:

day lets,
'do not disturb' notices on the doors,
late risers,
extra departures,
late departures,
food in rooms,
laundering in rooms,
rooms let for changing,
advances made to her,
requests from strangers for rooms to be opened.

There are instances when a room has been used for a day let, eg a business meeting during the day, and it requires a second service. This entails getting in touch with floor service to remove their equipment and making the room and bathroom ready for use that night.

Since the housekeeping wages bill accounts for a large part of the cleaning costs it needs to be kept as low as possible and the housekeeper has to employ only the minimum number of staff. In some places it has been found more practicable to employ part-time staff (women who work four to five hours in the morning and possibly others who work two to three hours in the evening), but it should be remembered that the casual hourly rate of pay is often higher than that for a full-time maid.

The room maid

The tendency today is for room maids to service a section of approximately 10 to 15 rooms with private bathrooms and the corridor, without any help from a cleaner; where there are few or no private bathrooms the room maids may be expected to service public bathrooms and WCs. The room maids' work is of great importance because it contributes to the comfort of guests and hence their impression of the hotel. It involves a knowledge and use of social and technical skills which should have been explained to them during their periods of training under the housekeeper or training officer. Room maids are responsible to the assistant housekeepers who, by supervising the maids and checking their rooms, ensure an accepted standard in the hotel.

In general, a maid's day consists of servicing each room to the required standard of the hotel and this includes bedmaking, coping with linen and general cleaning (see pages 21–7 for orders of work). All linen will be changed on a departure and in an occupied room according to house custom. In a luxury hotel this will be every day and in a less expensive hotel it will be less frequent (the towels are sometimes changed more often than the bedlinen), but when resheeting the changing of both sheets and both pillow slips (four on a double bed) is normal practice.

Soiled linen may be:

a) changed directly over the linen room counter by the maid, or houseporter;
b) bundled and taken to the linen room by the maid, house or linen porter at a set time each day and the clean linen collected or returned later;
c) despatched down a linen chute and the floor stock of clean linen made up later in the day by the house or linen porter;
d) collected frequently from the corridors by the linen porter and the clean linen returned later in the day.

Punctuality, neat appearance, courtesy and the anticipation of the needs of others are essential qualities in the maid. She needs to work quickly and efficiently following the housekeeper's instructions regarding methods of work and cleaning programmes. As well as this she needs to show some initiative in order that vacated rooms are serviced before occupied ones, that guests' requests for their rooms to be serviced at a certain time are met and that economies are practised with regard to time, labour and the use of materials.

She needs to work quietly and tidily causing as little inconvenience to the guest as possible. A maid in an hotel is usually allowed to turn the radio on low in the room in which she is working and normally she works with the door open. This, however, does present problems because the maid may be taken unawares by the guest or an intruder entering the room, particularly if the vacuum cleaner is working and so some housekeepers prefer the maid to have the door closed. The handling of guests' and hotel property should be such that no damage is done to the articles and nothing is thrown away without the certainty that it is rubbish. The maid should be fully aware of the need for safety and security within the department, avoiding any action which may lead to an accident or a fire hazard. She should understand the necessity for the strict control of keys and she should:

sign for them at the beginning of duty;
keep them on her person (never lend them);
never use them to let a stranger into a room;
always hand them in at the end of a duty period according to house custom;
hand in to the housekeeper any keys found left in doors or lying around;
never open communicating doors;
not open more than one door at a time.

To help the housekeeper and for the smoother running of the hotel the maid should realize the importance of reporting promptly such things as:

room occupancy ie state of the rooms and number of sleepers;
lost property;
any 'do not disturb' notices not removed by a certain time according to house custom;
light luggage;
missing or damaged articles;
anything in need of repair;
anything of a suspicious nature;
illness of a guest;
accidents;
hazards which may lead to accidents;
anything for which a charge should be made on a guest's bill eg early morning teas, dogs in rooms, etc;
requests or complaints of guests;
inability to enter a room for servicing;
presence of smells, syringes etc, indicating use of drugs;
'walk-outs' and unexpected departures;
unattended bags or parcels.

In order to do her work satisfactorily the maid needs to be told of relevant details for her section as early as possible, eg departures or requirements for new guests. This information will be given to her by the assistant housekeeper as soon as it has been received from the receptionist, initially by means of the Arrivals and Departures list and later by telephone or other means of communication between the housekeeper and receptionist.

The hours that a room maid may work vary a great deal; for some it may be a straight shift of eight hours for 5 days a week, and assuming the maids serve early morning teas they may work 7 am–3.30 pm. If the maids do not

Duty roster for 5 maids servicing 60 rooms from 8.00 am–9.00 pm

Maids	Mon.	Tues.	Wed.	Thurs.	Fri.	Sat.	Sun.	Wkly. hrs.
A	8–4	8–4	8–4	8–2 5–9	8–4	DO	DO	39
B	DO	DO	8–2 5–9	8–4	8–4	8–4	8–2 5–9	40½
C	8–4	8–4	DO	DO	8–2 5–9	8–4	8–4	39
D	8–2 5–9	8–4	8–4	8–4	DO	DO	8–4	39
E	DO	8–2 5–9	8–4	8–4	8–4	8–2 5–9	DO	40½

Mealtimes:
Lunch   12–12.30 pm or 12.30–1.00 pm
Supper   6.30–7.00 pm or 7.00–7.30 pm
8–4 =   8 hrs less ½ hr   = 7½ hours
$\frac{8-2}{5-9}$ =   10 hrs less 1 hr   = 9 hours (13 hours spreadover)

serve early morning teas then they might not come on until 8 or 8.30 am.

Where there may be 'turning down' and other duties to be done in the evening, separate part-time maids might be engaged for such hours as may be necessary, eg 5–9 pm, 6–10 pm. There would be fewer maids than in the morning as they would be expected to service larger sections. Many hotels now do not 'turn down' and in these places either one or two maids may be required in the evening to service late and extra departures and carry out any other necessary jobs or no maids will be on at all. Some hotels cannot implement straight shifts and split duties may still be necessary. In these cases a maid may work three or four straight shifts and one or two split duties per week.

During the course of her work the room maid may have half an hour for breakfast, and a mid-morning break of 15–20 minutes is taken in the staff canteen or her service room; lunch will probably be 12–12.30 pm or 12.30–1 pm. At arranged times during the day, according to house custom, the maid will deal with soiled linen and replenish her cleaning stores and stock of clean linen.

The first duty is to report to the duty housekeeper, collect her keys and obtain relevant information regarding her section. Her last job is locking her cupboards and service room doors and handing in her keys to the duty house-keeper as she reports off duty.

An example of a duty roster is given above for five maids with no reliefs servicing approximately 60 rooms; it is assumed that most business is carried on from Monday to Friday and so the weekends will be quiet periods. The coverage by the maids will be 8.00 am–9.00 pm with a housekeeper available from 4.00–5.00 pm when there are no maids on duty. Each maid has two days off a week and on a day when one maid is off duty the remaining four maids may service a split section in addition to their own and at the weekends, because of lower occupancy, three maids cover 60 rooms.

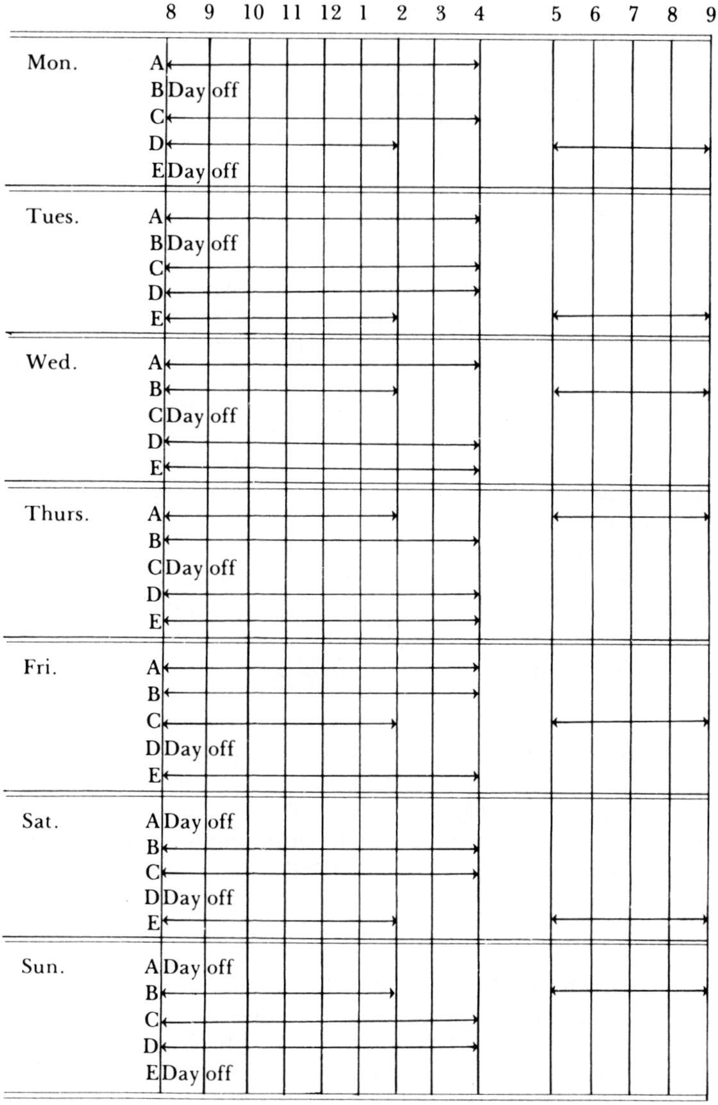

| | | 8 | 9 | 10 | 11 | 12 | 1 | 2 | 3 | 4 | | 5 | 6 | 7 | 8 | 9 |

Mon.
A
B Day off
C
D
E Day off

Tues.
A
B Day off
C
D
E

Wed.
A
B
C Day off
D
E

Thurs.
A
B
C Day off
D
E

Fri.
A
B
C
D Day off
E

Sat.
A Day off
B
C
D Day off
E

Sun.
A Day off
B
C
D
E Day off

Figure 3.1 Alternative method of setting out the same roster

In the hotel taken as an example, maids do not serve early morning tea but tea and coffee making facilities will be provided in each room and there is no need for maids to come on duty before 8.00 am. Some evening service will be provided for guests and so some split duties and overtime would appear to be unavoidable with this coverage and number of maids. To make an even share of duties the roster should rotate every five weeks.

A maid is entitled to be paid overtime if she works more than the statutory maximum number of hours, eg 39 hours a week depending on wages paid, and may also be entitled to extra money for spreadover, ie when work spreads over more than 12 hours. According to the Catering Wages Order, maids whose work is spread over more than 12 hours are entitled to extra pay, unless their current wage already exceeds the regulation minimum plus the extra allowances. Arrangements for a five-day week are made where possible, in some cases the two days off being taken together and in others not.

In some hotels maids live in; in others they live in a hostel or house provided for them; while in others they are non-resident. But in all cases they have their meals in the hotel, either in the staff canteen or their service room when on duty.

In the past it was customary for a room maid to wear a print dress, a large white apron and mob cap in the mornings, and a dark dress, a head band and frilly apron in the evenings. Now, with the advent of easy care materials, overalls have replaced this uniform and in some cases maids have to launder their own. Caps are not worn so frequently now, but maids are still required to wear stockings and sensible shoes.

Some maids make reliable assistant housekeepers but it is better that promotion is not within the same establishment, as it would be unwise to make a maid an assistant housekeeper over those with whom she has been working on an equal footing.

Staffmaids

In some hotels there is a staffmaid or cleaner who services the rooms of assistant housekeepers and other living-in staff. If staffmaids look after maids' rooms they are more likely to report a lack of hygiene and in some cases 'livestock'. They usually work a straight shift, and may be part-time, and are often considered when a room maid's job falls vacant.

Cleaners

The first job for cleaners will be to clean offices, public rooms and ladies' cloakrooms and possibly later they will clean corridors and stairs and carry out such other work as is required. Cleaners are usually part-time, always live out, have their meals in the staff canteen when on duty, and usually they wear coloured overalls provided by the hotel. Their hours will vary according to the work they have to do. Some hotels use contract cleaners (see page 50).

Cloakroom attendant

In an hotel which has many functions and many non-resident guests, it is usual to have someone on duty in a ladies' powder room, during lunch and dinner periods, who attends to the requirements of the guests, guards their belongings and keeps the powder room neat and tidy. The initial cleaning of the powder room will have been done by a cleaner.

If there is a need for the cloakroom attendant to be on duty in the powder room between 12–3 pm, and again from 6–11 pm this will be a full-time job. Often, however, the powder room is cleaned by cleaners in the morning, checked several times during the day by an assistant housekeeper and only 'manned' when required. Thus, the cloakroom attendant may be part-time, or may be a linen room maid or a staffmaid, who does not normally work in the evenings and so will take on this work as an extra. She wears either a dark dress and a frilly apron, or a white overall.

Note Cloakrooms for men are looked after by porters belonging to the uniformed staff; in addition to the normal fittings and fixtures, male cloakrooms may contain urinals, usually of the stall type.

Houseporters

In the majority of hotels the houseporter is the only man on the housekeeping staff. He may start work at 7 am and do a straight day working until 4 pm with appropriate meal breaks, taken in the staff canteen, working approximately a forty-hour week.

The houseporter's duties will vary from one hotel to another but, as a rule, if there are coal fires to be dealt with, these will be his first job; on the other hand, he may help with the cleaning of the foyer and public rooms, do the heavier vacuum cleaning, the shaking of door mats, the cleaning of lamps and shades and possibly dusting large mirrors. He keeps the 'tools' for his work in a cupboard allotted to him.

After his breakfast half hour there will be certain jobs that he will carry out each day, and others only when requested, and circumstances will dictate if and when he does the following work:

replenish cleaning stores according to house custom;
clean brasses, eg stair rods, and fire fighting equipment;
move furniture, eg cots, bed boards;
spot clean and maybe shampoo carpets;
take linen to and from the floors;
empty rubbish;
help maids with the moving of heavy furniture and the cleaning of high ledges and fitments;
take down and rehang curtains;
keep fire buckets filled with sand or water;
carry coal.

The houseporter usually wears a denim or a buff-coloured drill coat or jacket and apron provided by the hotel.

Valets

Valeting is only carried out on the premises of first-class hotels. It involves looking after the male guests; sponging, pressing and doing minor repairs to clothes; cleaning shoes; parcelling personal laundry and dry cleaning; moving guests' belongings when changes of rooms are necessary; unpacking

Figure 3.2   Valet

and packing for the guests. In the past a full-time valet was not on the house-keeping staff; he was either self-employed or on the uniformed staff. However, in newer hotels he may now be a member of the housekeeping staff because many of his duties, eg moving guests' belongings, are the direct concern of the housekeeper.

In many instances a valet is not fully employed with valeting alone and a valet-porter will combine the work of a valet with the less dirty jobs of the houseporter. He is a member of the housekeeping staff, so he may move furniture, transport linen, clean ledges and high lights, etc, but he must retain his neat appearance as he is always on call to give personal service to guests.

The valet has or shares a service room complete with iron and board, paper and string, needles and cotton, shoe cleaning necessities, and he usually keeps spare pyjamas, studs, braces, shoe laces, black ties, electric razors etc, in case of emergencies.

A valet or valet-porter wears formal uniform trousers and an alpaca waist-coat with sleeves but in some hotels he wears a special uniform. In a luxury hotel the coverage of the valets is similar to that of the room maids, possibly 7 or 8 am to 10 or 11 pm.

Assistant housekeepers

The assistant housekeepers, floor housekeepers or floor supervisors are those who supervise and check the cleaning of the guests' rooms, public rooms, offices and ladies' cloakroom. They supervise the room maids, cleaners and houseporters and check their work. In addition they have work delegated to them by the housekeeper, such as:

keeping record books;
compiling maids' rosters and holiday lists;
training of the maids on the job;

supervision of the stores and linen room;
despatch and receipt of the dry cleaning articles;
floral arrangements.

Where there is a number of assistant housekeepers there will be a senior or first assistant, who will be deputy to the head or executive housekeeper and whose duties will include checking some rooms especially VIP rooms and rooms OOO (out of order). Her delegated duties may be the allocation of maids, the compiling of duty rosters, holiday lists and details of hours worked for wage calculations. As far as possible the deputy's hours will be 9.00 am–5.00 pm but she may have to take a turn with weekend and other duties.

The assistant housekeepers' hours will vary considerably. Split duties are avoided where possible and they may work for example 7.00 am–4.00 pm; 9.00 am–5.00 pm; or 2.00–10.00 pm, 5 days a week. Minimum wages and holidays and maximum working hours are laid down by the Catering Wages Order. The assistant housekeepers may live on or off the premises and in the past they wore black dresses of their choice, but now there is more freedom of colour although in some hotels a uniform is provided.

In all aspects of her work, the assistant housekeeper must be aware of company, 'house' and departmental policies and it is better if she has had some period of induction before starting 'on the job'. No two hotels or house-keeping departments are run exactly alike so even an assistant housekeeper with experience has things to learn concerning the work of a particular 'house' when she is newly appointed.

It has been suggested that there should be one assistant housekeeper for every 50 rooms but as the assistants are never all on duty at the same time, due to rest days, shifts etc, the actual number of rooms that one assistant is responsible for in a large hotel may be nearer 100 or even more.

The first job of an assistant housekeeper on early morning duty (the duty housekeeper) will be to check in the maids and cleaners, issue keys, early call and morning tea lists and other relevant information, and supervise and check the cleaners and houseporters dealing with the public rooms and offices. The final jobs of a duty housekeeper in the evening will be to check the return of all keys, lock them away according to house custom, and to make sure that all safety and security regulations have been complied with on the 'floors'.

At all times the assistant housekeeper carries a notebook, pencil and a master key, which opens all bedroom doors, and she may be dismissed should she lose this key. After breakfast she goes to her floor, checks on her maids and collects information for the housekeeper's report (room occupancy list see p. 300); as the assistant housekeeper is responsible for seeing that the work of her floor or floors is completed on time, she may have to ask maids to service extra rooms for extra money or in exchange for vacant rooms, because of sickness, holidays or days off. Where possible she quickly checks vacated rooms before the maid enters the room, making notes regarding lost property, missing or damaged articles and jobs for main-tenance, etc. For some of this information she relies on her maids because inevitably they will get into some rooms before she can.

From time to time the assistant housekeeper will return to the housekeeper's office to enter details from her notes into the appropriate books and to collect any further information regarding her floor, eg extra and late departures, moves, etc. Duplicate books are ideal for the reporting of maintenance and for information to reception, as the top copy goes to the relevant department and the other copy is kept as a record.

When the maids have completed servicing a vacated room the assistant housekeeper thoroughly checks it, prior to passing it to reception as a ready room (ie ready for letting). She notes any maintenance not yet done and oversights on the part of the maid, whom she will bring back to finish the job.

Inspections

A suggested method for an assistant housekeeper to use when checking (or inspecting) a bedroom with private bathroom is as follows:

1 In a clockwise or anti-clockwise direction check systematically everything on or touching the walls, eg:
  door – top, architrave and lock;
  electric light switches, telephone, radio and television;
  bed – making up, clean linen, castors, headboard etc, bed light, bedside
    table and contents;
  dressing table – clean and re-lined drawers, drawer stops, mirror,
    disclaimer notice, laundry lists, etc;
  wardrobe – clean shelves, rail, coat hangers, mirror, hinge and lock of
    door;
  window – sill, sashcords, catches, curtains, pelmet and runners;
  radiators;
  air conditioning;
  luggage rack;
  pictures.

2 Check free standing furniture, eg:
  armchair – upholstery, castors;
  standard lamp;
  occasional table and ashtrays;
  dressing table stool or chair;
  waste-paper basket.

3 Check ceiling and floor, eg:
  ceiling and central lights for soiling and cobwebs;
  floor – carpet, surround, etc.

4 General surveyance of the room.

5 Private bathroom, eg:
  door – top, architrave, lock and hook;
  light switch;
  vanitory unit – mirror, tiles, toothglass, light, razor point, taps, basin
    (underside, overflow, plug and waste hole), new soap, clean towels and
    guests' supplies;
  WC – bowl inside and out, underside of seat and lid, toilet paper;

ashtray and sanibin;

bath and shower – tiles, chrome fittings, soap recesses, the tub (overflow, plug and waste hole), towel rack, shower curtain and rail, new soap, bath mat and non slip rubber mat according to house custom;

'give away' toiletries;

ceiling and floor;

general surveyance of room.

Note This assumes that the bathroom is internal; should there be a window, then sill, catches, curtains and runners should be checked. In more luxurious hotels there may be a bidet which would be checked in a similar manner to the WC.

See also Figs 3.3–3.7 on pages 45–9.

In the course of the assistant housekeeper's checking of a ready room she may come across maintenance work which had previously been overlooked (or not done) and she should make a note of it and report it according to house custom. If there are serious discrepancies in the maid's work she should be brought back to put them right; but for minor faults she may be told about them and reminded to avoid them in the future.

The assistant housekeeper passes ready rooms to reception as soon as she is able, checking vacant rooms as early in the day as possible and occupied ones when she can. Inspection needs training and experience, as thoroughness with speed must be the aim of every housekeeper. The satisfaction of the guests with regard to the appearance and cleanliness of their rooms rests almost entirely with the assistant housekeepers, as it is they who motivate and supervise the maids who actually carry out the work of servicing the rooms, although of course the housekeeper has ultimate responsibility.

At any time during the day the assistant housekeeper may have interruptions from guests making requests, or perhaps complaints, and this all takes time. She impresses upon her maids the need for courtesy to the guests and a neat and tidy appearance both in themselves and in their work and also the need for fire precautions, the care of keys and other aspects of safety and security. She should be friendly towards her maids without undue familiarity, firm, fair and constructively critical and should give praise where it is due.

In hotels where there is a housekeeper with only one or two assistants an assistant housekeeper may be on duty alone and will deputize for the housekeeper when necessary, she is then sometimes called the duty housekeeper. In such hotels, the housekeeper herself may at times be the only one on duty and she is therefore much more involved in the day-to-day routine of the checking of rooms etc than the head or executive housekeeper in a large hotel.

In many small hotels there are no assistant housekeepers and a general assistant may combine housekeeping with many other duties, eg reception and the bar.

The aim of the assistant is to become a housekeeper and any further qualifications that she can acquire, such as a foreign language or HCTB certificates (eg Instructor trainers' course), will be a help in her present work and towards her promotion.

The housekeeper

As the housekeeper has overall charge of the department she has to be sure that standards are maintained throughout. Thus, during the day she will visit as many areas as possible, using her expertise to notice badly hanging curtains, poorly arranged furniture, soiled carpet or upholstery, a badly made bed, a stained bath, etc.

It is usual for there to be a housekeeper's office where the *housekeeper* may discuss the affairs of the day with her assistants and sit and do necessary paper work, such as maintenance reports, rosters, records, etc.

The *housekeeper* should not get so immersed in paper work that she stays in her office all day; she should be seen about the house observing people and things, in other words have 'time to stand and stare' and in order to be contacted she should carry a 'bleep' (see p. 88). To gain this time, she has to delegate work to her assistants. However, it must be emphasized that delegation does not mean the loss of responsibility, but rather the entrusting of it to a deputy.

In supervising and controlling her staff on duty, the *housekeeper* must also be concerned with their personal hygiene, and so she endeavours to make them realize the importance of the cleanliness of all parts of the body, the necessity for suitable clean clothing, including shoes and stockings, and for a neat and tidy appearance. The *housekeeper's* attitude sets the tone of the whole department and the staff take their cue from her, so she should at all times be an example that they can look up to and respect. Courtesy and good manners are essential and in the hotel all guests should be addressed as 'Sir' or 'Madam'.

A *housekeeper* needs to provide up-to-date equipment and methods of work to save the time and energy of the staff. She needs to be a good disciplinarian and be firm regarding the rules that she makes. In order to keep her staff she must realize there will be times when she has to give way and in doing so she must be just and fair and avoid favouritism. When compiling a duty roster, it should be made acceptable to the hotel and the individual members of staff. As far as the hotel is concerned, statutory regulations have to be met and the wage bill kept as low as possible while standards are maintained. For the work to be carried out efficiently, and to make a fair work load, there must be the maximum staff at peak periods and adequate staff at other times. There should also be an even share of duty hours (no unfair share of early mornings or late evenings), as few split duties as possible and early notification of changes (see p. 84).

A good *housekeeper* will always have the welfare of her staff at heart and will be prepared to stand up for them at all times (see pp. 84–5).

For further details of the housekeeper's work see Chapters 1 and 6.

Staff duties in non-commercial (welfare) establishments

In most establishments other than hotels the staffing is for cleaning and there is little personal service. The hours worked by the cleaners will depend on the areas to be maintained by the housekeeping department and the number of staff employed to undertake that work. As many of the staff will be part-time cleaners, their hours will determine the amount of work which can be assigned to each employee.

In various types of *hostels* and *'homes'* little work can be done before 9 am so cleaners may work from 9 am–1 pm, 9 am–3 pm or even 9.30 am–3.30 pm, and so the amount of work assigned to them will vary. In some residential establishments public rooms need to be cleaned before breakfast and so some cleaners start earlier than others.

In any of these residential establishments cleaners are hindered by residents being in their rooms and it is generally recognized that there will be days when the very minimum amount of work will be done in a particular room, due to it being occupied. In students' hostels this applies especially at weekends and in many cases there will be no cleaners in on Saturday or Sunday for the cleaning of the study bedrooms. Occasionally there may be a few full-time maids – possibly resident – whose duty rosters will be arranged so that they may clean public rooms, vacated guest rooms where these are provided, and similar jobs which become necessary. Thorough cleaning of students' hostels is normally carried out each vacation when the rooms are empty but difficulties do arise because, for the economic running of the hostels, facilities have to be available for conferences and holiday bookings during the major part of the vacations. Careful planning should be given to the letting of the hostels for such bookings to enable the extra cleaning, including annual cleaning with possible redecoration, to be carried out. In other establishments, thorough and annual cleaning will be carried on throughout the year as is convenient to the particular place.

Some of these hostels and hospitals etc, are large and may cover a wide area. Problems may arise when cleaners are required to clock in and there is an inevitable time lag between clocking in, obtaining their keys and getting to their place of work.

Cleaners

Cleaners in various types of hostels and halls of residence (including nurses' homes etc) may service a section of 10–20 rooms with or without wash basins, utility room, bathrooms, WCs, showers and corridor. Assuming the cleaner starts work at approximately 9 am she will first clean the rooms of her section, completing each room preferably on one visit. Residents often have expensive TVs, videos etc, and so for security reasons block cleaning, certainly in the NHS, is not recommended.

In many instances the beds will be made by the residents and the cleaners will only make them on clean sheet day. In hospitals all residents (ie not patients) change their own bed linen on clean sheet day and rooms usually receive a weekly clean. Sometimes residents clean their own rooms with equipment provided by the hospital and when this happens domestic staff will only clean public areas, eg public sitting rooms, corridors, cloakrooms, bathrooms and WCs.

Where room cleaning is carried out, in many establishments it is not done at the weekends, so the rooms on Mondays will need a little more attention and Mondays should be avoided as the day when the linen is changed. It is usual to send the bottom sheet and one pillowslip to the laundry each week and so the top sheet becomes the bottom sheet and a clean one is put on the top. Thus one clean sheet and one clean pillowslip are provided each week. (The residents may provide their own towels and, less frequently, bed linen.)

A mid-morning break of about 15 minutes is usual and it is possible that about this time on given days the cleaner will collect her cleaning stores according to house custom. If the cleaner works after 1 pm she will be given 30 minutes–1 hour break for lunch, which she will probably take between 12 noon and 2 pm. After her lunch she continues her work and before going off duty she deals with her equipment and hands in her key.

If there are areas, eg offices, which require cleaning before breakfast they will either be dealt with by a living-in maid or a cleaner who comes in especially early. Other areas, eg entrance halls, common rooms, dining rooms, etc will be cleaned at a time convenient to the house.

The daily routine for cleaners in places let during vacations for conferences, summer schools etc, may be considerably different from that during term time as guests may require more service, eg bedmaking. The rooms may be let for comparatively short periods and there may be a number of departures on one day with little time for cleaning between one let and the next.

It is usual for a cleaner to wear an overall, which may or may not be provided by the establishment.

Male domestics (porters)

Male domestics (porters) in various types of hostels and halls of residence can be likened to houseporters in an hotel and their duties may include dealing with rubbish, moving furniture, cleaning carpets, cleaning fire escapes, transporting linen, etc. They may wear a denim coat or dungarees and will work such hours as are required by the establishment.

Supervisors

In large establishments where there are many cleaners and large areas to cover, there may be supervisors who are responsible for the day-to-day control of a number of cleaners in a particular area.

Junior domestic bursars/assistant housekeepers

The main work of a junior domestic bursar is the supervision of the cleaners and the checking of the rooms. During this time she makes out maintenance lists according to house custom and checks that previous maintenance work has been carried out. Any urgent repairs will be immediately reported to the appropriate person. In addition to this work she fits in any work delegated to her by the domestic bursar/housekeeper.

One assistant will be on duty just before the cleaners start work in order to check them in, issue keys, pass on information and rearrange the work of absentees. Assistants will normally work some split shifts as it is usual to have one of them on call in the evening.

Housekeeping staff in hospitals

Until comparatively recently the domestic services of hospitals were the responsibility of the nursing administration. As it was realized that nurses are trained to nurse, and should not spend valuable time on work that does not require nursing training, these duties became the prerogative of lay staff trained in domestic management. This encompasses all the new developments in up-to-date cleaning equipment and techniques required for the proper cleaning of hospital premises and other duties that do not require nursing skills, eg linen services, refuse control etc (see p. 4). This led to the introduction of various forms of ward housekeeping schemes staffed by the domestic services department but under the local control of the ward sisters, or nurses in charge of the wards. A housekeeper or domestic supervisor may lead the team and is responsible for supervising some, and co-ordinating other, housekeeping services in the ward or nursing unit to which she is attached. Schedules of work are agreed with the ward sister or nurse in charge and the deployment of staff to the unit and the assurance of their technical competence remains the responsibility of the Domestic Services Manager.

The Domestic Services Manager is responsible for the housekeeping of all parts of the hospital, including staff residences. This involves vast floor areas which are in use twenty-four hours a day every day of the year. Special areas, such as the intensive care unit, theatres, renal unit, transplant unit and the premature baby unit are also included and accurate planning of work schedules is essential because the availability of these areas is limited and they often have to be cleaned outside normal hours. In isolation and high-risk areas the exact tasks carried out by the domestic services staff will depend on the unit concerned and it is often necessary to give training in the special methods and precautions required.

With the advent of competitive tendering the contract manager has similar responsibilities but in some hospitals there has been a reduction in some of the non-nursing duties previously undertaken by domestic staff in the wards. In any one establishment, however, only one specification of work is drawn up and care must be taken to ensure that the contents of the tender documents are equally applicable to in-house and contract cleaning firm tenderers.

A result of some NHS staff becoming contract managers and vice versa (contract managers becoming domestic services managers) has been the growth of a better understanding of the work required in hospitals, whether by in-house or contract cleaning staff.

Problems which have to be overcome and which are not so likely to be met in other establishments include the following:

  routines need to be planned so that they do not inconvenience doctors, nurses or visitors;
  cleaning has to dovetail with nursing procedures, eg bedmaking before vacuum cleaning;
  a greater need for flexibility and the ability to adjust to interruptions;
  methods and equipment require special attention so as to prevent cross-infection;
  the noise factor is of more importance than elsewhere;

greater variations in different work methods and standards in different areas, eg operating theatres, wards and visitors' lounge (areas of high, medium and low risk);

the need for check cleaning, that is the frequent re-doing of cleaning tasks in heavy user areas, eg mopping of floors, cleaning of WCs and wash hand basins;

the irregular hours of medical and nursing staff, especially in regard to the servicing of staff accommodation.

Domestic assistants, ward orderlies and central cleaning teams

These grades are employed to relieve nurses of non-nursing duties in patient areas and to clean all other parts of the hospital. They normally work under incentive bonus scheme conditions.

Domestic supervisors/ward housekeepers

The supervisors or ward housekeepers are first line managers who supervise the work of the domestic assistants and ward orderlies in a given area. They are responsible to the senior housekeeper or assistant domestic services manager for maintaining the correct methods of work and standards. They are responsible for the training and allocation of staff in their areas to ensure that the correct methods and standards are maintained. They liaise with the ward sisters/charge nurses on the work within the wards.

Assistant domestic services managers and senior housekeepers

Depending on the size of the hospital a domestic services manager is supported by a number of assistant domestic services managers or senior housekeepers. When a hospital is too small to justify the appointment of a domestic services manager an assistant domestic services manager or senior housekeeper will be responsible to the domestic services manager or Unit Administrator for the supervision of domestic services.

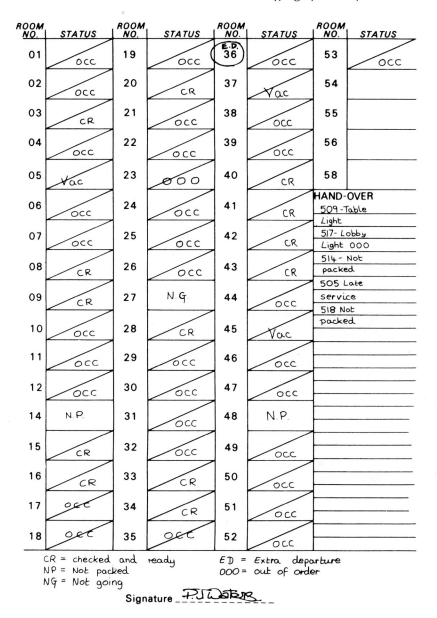

ROOM NO.	STATUS	ROOM NO.	STATUS	ROOM NO.	STATUS	ROOM NO.	STATUS
01	OCC	19	OCC	E.D. 36	OCC	53	OCC
02	OCC	20	CR	37	Vac	54	
03	CR	21	OCC	38	OCC	55	
04	OCC	22	OCC	39	OCC	56	
05	Vac	23	OOO	40	CR	58	
06	OCC	24	OCC	41	CR	HAND-OVER 509 -Table Light	
07	OCC	25	OCC	42	CR	517- Lobby Light OOO	
08	CR	26	OCC	43	CR	514- Not packed	
09	CR	27	N.G.	44	OCC	505 Late service	
						518 Not packed	
10	OCC	28	CR	45	Vac		
11	OCC	29	OCC	46	OCC		
12	OCC	30	OCC	47	OCC		
14	N.P.	31	OCC	48	N.P.		
15	CR	32	OCC	49	OCC		
16	CR	33	CR	50	OCC		
17	OCC	34	CR	51	OCC		
18	OCC	35	OCC	52	OCC		

CR = checked and ready   ED = Extra departure
NP = Not packed   OOO = out of order
NG = Not going

Signature _P.J.Webb_

Figure 3.3   Specimen floor checklist (hotels)

BEDROOM INSPECTION REPORT

Miss Johnson

	581 VAC	571 VAC	574 OCC	583 DEP
Ashtrays	✓	✓	✓	✓
Bed making	✓	✓	✓	✓
Bed wheels	✓	dusty	✓	✓
Bed head	✓		✓	✓
Bed sides	slight dust	✓	✓	✓
Bed unit	✓	✓	✓	✓
Blotter contents	✓	stained	✓	✓
Carpets	not hoovered	✓	not hoovered	✓
Carpet edges	neglected	neglected	✓	neglected
Chair arms	✓	✓	✓	✓
Chair dressing table	✓	✓	✓	✓
Coat hangers	✓	✓	✓	not enough supplied
Dressing table unit	✓	✓	✓	✓
Drawers	✓	✓	✓	not checked
Door tops	✓	✓	✓	✓
Door Comm.				
Door jambs	✓	✓	✓	✓
Furniture	✓	✓	✓	✓
Furniture fronts	✓	✓	✓	✓
Furniture legs	✓	✓	✓	✓
Lobby skirting	✓	✓	✓	✓
Lobby carpet	not hoovered	not hoovered	not hoovered	not hoovered
Lights bedside	✓	✓	✓	✓
Lights dressing table	✓	slight dust	✓	✓
Lights standard lamp	✓	✓	✓	✓
Laundry bags (2)	✓	✓	✓	only 1
Mirrors	✓	✓	✓	
Pictures	✓	✓	top of picture dusty	✓
Skirtings	✓	✓	✓	✓
Tables Coffee	stained	✓	✓	✓
Tables bedside	✓	✓	✓	✓
Telephones	✓	✓	✓	✓
Ventilator	✓	✓	✓	✓
Wardrobe shelves	✓	✓	✓	good
Wardrobe floor	✓	✓	✓	✓
Wardrobe rack	✓	✓	✓	✓
Wardrobe rail	✓	✓	✓	✓
Window ledges	✓	✓	✓	good
Wastepaper bin	✓	ash still in bin	✓	✓

Housekeeper's signature A J Wilson ....

Figure 3.4   Bedroom inspection report (hotels)

<u>GENERAL MAINTENANCE REPORT</u>

Dept: Housekeeping   *Floor* 5ᵗʰ   *Signature* Sue Woolt   *Date* 29·3·

ITEM	ROOM NUMBER							
CONSTRUCTION								
Bedroom curtains–a) track b) runners								
Bath–a) tap b) slow waste c) plug off d) stained	501ª							
e) chipped f) seal								
Basin–a) tap b) slow waste c) plug off d)stained								
e) cracked f) seal								
Bathroom flooring –a) marked b) torn c) loose	584ª	523ª	511ᶜ					
Chairs–a) broken b) dirty c)torn								
Carpet–a) minor repair b) spot c) shampoo	503ᵇ							
Dressing table drawers–a)broken b)handles off								
Dado rail–a) off b) torn								
Door closer–a) not working properly								
Furniture–a) cigarette burns								
Loose/defective fittings–a) hooks b) towel rail								
c) door stops d) bottle openers e) drip dries	571ᶜ	588ᴰ						
f) tissue dispenser								
Mirror–a) cracked b) spotted c) domeheads	564ª							
Paintwork–a) chipped b) dirty								
Polishing–a) door b) furniture								
Shower curtain –a) track b) runners								
Threshold strip to bathroom –a) loose b) off								
Toilet seat– a) loose b) chipped								
Wall tiles–a) cracked								
W.C.–a) stained b)cracked								
Wardrobe–a)catches b)shelf c)hinges d)hanging rail								
Window–a)not opening b)not closing c)not fastening								
CORRIDORS								
Ceiling sections–a) missing b) dirty								
Signs–a) loose b) broken c) missing								
Wall covering–a) torn b) dirty								
ELECTRICIANS								
Bedhead console switches–a) knobs loose b) missing								
Bathroom extract–a) dirty b) not working								
Lobby light–a) not working								
Lampshades–a) chipped b) cracked								
Room status indicator–a) not working								
ENGINEERS								
Extract –a) dirty								
Room 529 Double lock not working								

Figure 3.5   General maintenance report

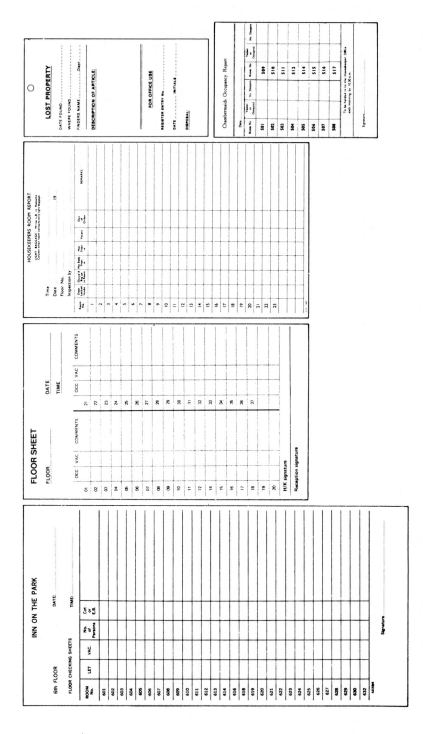

Figure 3.6 Specimen record sheets (hotels)

Date

WARD/DEPARTMENT	STANDARD			Action taken by Supervisor
	Good	*Fair*	*Poor*	
Carpets				
Vacuum cleaned				
Spot free				
Floors				
Vacuum cleaned				
Damp mopped				
Dressed and buffed				
Furniture and fittings				
Damp mopped				
Polished				
Upholstery vacuumed				
Sinks, basin				
Spot free				
Inside				
Outside				
Underneath				
Taps				
Plughole, plug. and chain				
Overflow and fitments				
Lavatories				
Inside				
Outside				
Underneath				
Backs				
Seat				
Chain				
Brush holder				
Baths				
Inside				
Outside				
Underneath				
Taps				
Plughole, plug and chain				
Overflow and fitments				
High cleaning				
Curtain rails				
Door ledges				
Tops of cupboards				
Ledges, pipes				
Low cleaning				
Corners				
Pipes				
Radiators				
Wheels				
Under furniture				
Rubbish				
Ashtrays				
Disposal bags and stands				
Incinerators				

Figure 3.7 Specimen checklist (hospitals)

4

CONTRACTS

Contracts may be made with some firms for the provision of certain services and for the hire of some articles. The housekeeper may consider carefully the advantage of having these contracts instead of:

finding, training and supervising in-house staff to undertake these services;

obtaining specialized equipment for infrequent jobs and buying articles.

Contract cleaning

Contract cleaning firms offer a wide range of services which will include:

complete cleaning programmes with all work and responsibility undertaken by the contractor;

regular, selected types of cleaning within an establishment to assist the existing housekeeping organization, eg night cleaning of entrance halls, washroom cleaning etc;

periodic services to assist the existing housekeeping organization, eg window cleaning, wall washing, descaling and disinfecting of sanitary fittings, carpet and upholstery cleaning etc.

At the time of writing the National Health Service has a system of competitive tendering for the domestic services in hospitals; an in-house tender may be put in along with those of contract cleaning firms and the tenders are given equal consideration.

In hotels, contractors are more usually employed for selected types of cleaning, both on a regular and periodic basis.

When choosing between contract and direct (in-house) labour, cost, service and convenience are important points to be considered.

Basically the decision between contract and direct labour must be made on the amount of money available. It has been suggested that a contractor must be *20–30 per cent more productive* than direct labour in order to provide an *equal service* at an *equal cost* and still get a fair profit.

With a good contractor it is possible that a higher level of cleanliness at the same cost may be obtained because new methods and more efficient equipment and materials are used. Materials will be bought in large quantities, and therefore more cheaply, and this may be reflected in the cost of the service. (Materials, however, represent only a small proportion – possibly 5 to 10 per cent – of the total cost of any cleaning process.)

The main advantages of contract labour to the client are:

there is no capital outlay for equipment so money is available for investment or other purposes;
there is no equipment lying idle (particularly specialized equipment);
there is no buying or hiring of specialized equipment;
the difficulty of finding, training, organizing and supervising the cleaning staff is passed to the contractor;
extra work may be carried out at certain times without increasing the basic staff;
the exact cost of cleaning is known for a given period.

In spite of these advantages there are many dissatisfied clients and many of them have reverted to direct labour.

Causes of dissatisfaction may include:

loss of flexibility to effect changes; the *housekeeper* no longer controls the operation;
loss of proprietary interest. The cleaners do not belong to, ie do not work for, the establishment and may not have the same pride in their work or job satisfaction;
problems regarding security;
problems regarding liaison and co-operation between departments;
deterioration in the quality of the work. One of the reasons for this may be the great growth of contract cleaning firms, resulting in cut-price tenders being offered and accepted (clients almost always accept the lowest price), which do not enable contractors to employ a sufficient number of employees or employees of the right calibre to do the job properly. Supervision is all important and the right supervisor essential; good workers should be the result of good supervision.

It is essential that the *housekeeper* provides the contractor with a detailed specification of the work required to be done, when it is to be done, how it is to be done etc, and from this the contractor draws up the contract.

Deterioration in the quality of work may not always be the contractor's fault but can result from a poorly worded, insufficiently detailed specification, with the housekeeper not realizing that the contract drawn up from it is incomplete in detail. Specifications for the contractor should be similar to ones drawn up for in-house labour doing the same jobs. Specifications should therefore be carefully worded and may cover the following points:

schedule of areas to be serviced and the frequency with which a job is done. This is important as the level of cleanliness depends on the time elapsing between successive cleaning processes, eg dusting of horizontal surfaces daily, and vertical weekly;
description of method, equipment and materials required;
hours during which work is to be done;

security requirements – sometimes all staff are 'vetted', and all specified rooms have to be kept locked;

provision of adequate supervisory requirements;

storage areas and provision of lockers and other accommodation for the staff;

cover for sickness and annual leave;

specified frequencies of inspections with the contractor;

necessity of complying with the Health and Safety at Work etc Act;

public and customer liability (insurance).

The contract itself will state the duration of contract, the price for the job (often expressed as cost per sq.m.) and whether an initial clean is required before the contract comes into operation. It should also give provision for a regular review of the specification and for information to be given to the client regarding operating costs.

Costs

Most contracts are agreed on a unit rate agreement. The *housekeeper* provides details of the area, the frequency of the job and asks for the cost. The reliable contractor measures and calculates the cost.

Man-hours = areas × time × frequency

To the cost of wages (operators and supervisors) he adds cost of equipment, agents and supplies, overheads and profit.

The contractor who wants the job may guess the lowest price and the contract invariably goes to the lowest bidder who may have to cut standards to cover his losses.

Far better than the contract agreed on a unit cost basis is one in which the contractor is paid for the costs of the job and given a fixed fee. In this case, there is no point in his cutting his cost because he has a guaranteed profit. A one-year contract basis is thought to be unsatisfactory; a three or five-year contract is better and the costs should escalate over the period.

With the 'costs plus fixed fee' type of contract, the *housekeeper* specifies staff and equipment etc. Specifications for a contractor should be the same as for direct or in-house labour and the *housekeeper* should compare the contractors' bids with her own in-house cleaning cost, eg:

400 hours per week at xp an hour	=	£400x per year
Supervision		y
Small equipment and supplies (cloths, etc)		z
Capital equipment (written off over 10 years)		
10 vacuum cleaners	£a	
1 polisher	£b	
1 scrubber	£c	
2 dustettes	£d	

$$£a + b + c + d = £m$$

Capital investment over 10 years $£\dfrac{m}{10}$

Cleaning agents $£\ n$

Total cleaning cost per year is

$$£\ 400x + y + z + \frac{m}{10} + n$$

This can then be expressed as cost per square metre if required.

Other contracts

Besides contract cleaning other contract arrangements may be made by the *housekeeper* for her department with:

a laundry – when a price is approved by weight or per flat article, provided a minimum number is sent;

a florist – when an agreement is made to provide floral arrangements for specific areas over a given period at a specified price;

various manufacturing firms for the –
servicing of equipment at stated intervals,
delivery of certain goods at stated times;

various hire firms –
linen (p. 133),
cleaning,
equipment, (p. 104),
furniture and furnishings (p. 258),
TV,
sanitary disposal services (p. 295),
in-house laundry equipment,
conference equipment,
dust control mats.

In the case of equipment, furniture and furnishings the contract is made for a given number of years, after which a new contract for new articles may be made or the old articles may be retained at a reduced price. *Leasing contracts* for hard furniture and upholstery in leather or expanded vinyl are generally written over a 5–7 year lease period, while those for soft furnishings are for a shorter period – generally three years.

At the end of the primary leasing period, when a new contract for new articles is written, it is usual for a 'trade-in' allowance to be given, equal to the residual value of the original items.

In all cases it is necessary to pay great attention to the terms of the lease and to use firms of good repute. However leasing does mean no capital outlay, budgeting is made easier because costs are known and it is possible to modernize immediately and pay for the modernization with future profits.

5

SECURITY, SAFETY AND FIRST AID

Security

Security is not the prerogative of any one person in an establishment; all staff should be security minded and report anything of a suspicious nature. Staff should realize the necessity of not giving information regarding internal matters to such persons as enquiry agents, newspaper reporters, etc.

Most large establishments, eg hotels and hospitals, have one or more security officers on their staff to prevent crime and to protect *guests* and their staff from such dangers as theft, bomb threat, fire or assault. In smaller hotels the responsibility for security will be the manager's, and in other establishments will be that of the manager's equivalent.

A security officer, often an ex-policeman, keeps in touch with other security officers and any information gained is shared among them. He should have sufficient seniority to command respect from the staff and to ensure that the necessary measures are carried out. The security officer moves inconspicuously among the *guests* and is responsible for arrangements regarding:

suspicious persons or behaviour;

keys, electronic locks and window locks;

bomb threats, fire precautions and the evacuation of the building if necessary;

inspection tours of the building to check for security hazards, suspicious objects and to deal with them accordingly;

the number of unlocked entrances and exits;

closed TV for identification of persons entering;

contractors and casual staff entering the building;

searching of staff bags and body searches;

adequate watch on the premises to prevent prostitution;

lost property procedures;

investigation of reports of *guests'* losses;

provision of safety deposit boxes in *guests'* rooms and 'peep' holes in the doors;

the safeguard of money when large amounts are being moved from place to place eg to and from the bank.

The security officer may or may not also be the safety officer and so may or may not be responsible for safety precautions.

Good hall porters, by experience, get to recognize people with a furtive air or remember those who have given trouble in the past. Head hall porters in an hotel may belong to an association through which they exchange information regarding undesirable characters.

Entrance halls of all establishments are vulnerable places. In large places, hotels, hostels, hospitals etc, there can at certain times be many people about and thieves and terrorists may take advantage of this, eg picking up unguarded articles or taking the opportunity of getting further into the building. In small establishments the entrance hall is often not 'manned' and, unless the door is locked, anyone may enter. The time-keeper keeps an eye on the back door and staff comings and goings, and at times may inspect parcels and cases according to house custom.

There should be as few unattended doors to the street as possible, and at night all outside doors, except fire doors which should only operate from the inside, should be locked and late staff should enter by the front door. Ground floor windows and french windows should have safety catches, and these should be firmly secured at night.

The *housekeeper* and her staff are about the building perhaps more than many other staff and must be aware of the ways in which they can be security minded. If a thief wants to get into a room, he may gain admittance by telling the maid he has:

a repair to carry out;
come to collect the television set or other articles;
flowers to deliver to a certain room;
forgotten his key.

Therefore, a maid should be instructed to keep a look out for, and report as soon as possible, any suspicious characters and be warned against opening doors for strangers; when such requests are made she must say that she cannot unlock the door, but will fetch the *housekeeper* who should check the name of the guest with reception. Maids should be instructed to lock all doors on leaving a room and to remove all keys left in doors immediately they are seen and hand them to a *housekeeper* who will return them according to house custom.

Keys

Since management has certain responsibilities for the safety of the *guests'* belongings, the proper care of keys is a very important aspect of security.

Individual heads of departments are responsible for all the keys in their areas and the housekeeper probably has control of more keys than any other departmental head. In an hotel there are grandmaster, master, submaster and individual room keys.

Grandmaster key

This key:

opens all doors and, in addition, double locks them against all other keys;

overrides the catch put on by the guest for privacy in the room – a precaution necessary in case of an emergency, eg illness or injury;

is used when access to a room has to be prevented, eg in the case of death;

is used when a guest leaves his belongings in his room and goes away for a night or two;

is used when a guest does not leave his key at reception and the guest needs to be seen by the manager for some special reason;

is held by the security officer, general manager, duty manager and sometimes by the housekeeper.

Master key

This key will:

open all rooms in the house,

be carried by assistant housekeepers and floor service waiters while on duty, and sometimes by maintenance staff.

Sub-master key

This key:

opens all rooms in a maid's section;

is signed for at the start of work;

is attached to a belt round the maid's waist and should never leave her person;

should never be lent to anyone;

is handed in when the maid goes off duty.

Last thing at night, the duty housekeeper checks the return of all keys and locks them away for the night.

It is generally written into staff terms of contract that the loss of any master key will lead to dismissal of the employee.

Individual room key

On being shown to a room a guest is given a key with a room number and the name of the hotel on the tag. Guests are asked to hand in their keys when they go out and the keys are then put on a key-board which should be out of view of passers-by as another security precaution. A key not on the key-board should indicate that the guest is in the hotel; this information can be useful in the case of an emergency.

The mastering of locks is necessary but it should be realized that when locks are mastered a certain amount of security is lost and great care should be taken in the allocation of keys to responsible persons, emphasizing the need for the utmost care in their protection and use at all times. It may be possible to zone rooms so that not all parts of the establishment are in jeopardy should a master key be lost.

Key thefts

The locks in hotels are generally spring operated mortice locks (see page 280) but to overcome the problem of key thefts, keyless lock systems have been devised. They are expensive but are in use in some hotels. They may be computerized but there is also a less expensive battery-powered lock system, with an infra-red device which reads the guests' cards. A disposable plastic card about the size of a conventional credit card replaces the normal metal room key. One of these cards is given to each guest on checking in. In the computerized system the card is coded by perforations at random from a pool of more than four billion potential codes available from a master computer console at the front desk. The random code is then transmitted electronically to the specific guest's room lock and only this particular card can open that room door. When the guest checks out, the code on that room is changed and a new guest will receive a new code on his new check-in card; the old card automatically becomes useless. Similar cards may be coded as master keys for the maids and housekeepers and changed at frequent intervals at reception.

Re-keying lock systems may also help to overcome the problem of key thefts and a new system allows the lock to be changed quickly and easily without having to take the tumbler apart, without removing the lock from the door and without entering the room. The lock is changed by using a special key which instantly changes the tumbler.

Lost property

It is general practice that any lost property found in rooms should be handed in to the *housekeeper's* office immediately (or other place according to house custom), and the appropriate details should be entered in a lost property book, after which the articles should be labelled and will usually be kept for a period of six months. Great tact should be exercised in dealing with lost property and it is advocated that *guests* are not notified of articles found in rooms unless they are still in the building. Precautions need to be taken to ensure that articles are only handed over to the rightful owner and not to any would-be claimant.

Valuables

Hotels have a *safe* or *safe deposit boxes* and notices are displayed asking guests not to leave valuables in their rooms but to have them locked away in the safe. Should a maid come into an occupied room and find valuables left there, she should inform the housekeeper who will deal with them according to house custom. It is less likely that guests will leave valuables about when *personal safes* are provided in their rooms. Peepholes which allow guests to see who is outside their door are considered standard security equipment in some hotels.

On being admitted to hospital unexpectedly a patient may have valuables or a large sum of money which should be put into safe keeping. Similarly on the death of a patient there may be articles to be kept until claimed by the next of kin, so suitable security arrangements need to be made in all places. In some establishments (eg college halls of residence) residents are advised to take out personal insurance against theft.

Other security measures

The *housekeeper* is responsible for the reporting of faulty window catches etc, and at night should ensure that all french windows and balcony doors are securely locked and that panic bars on fire exit doors are adjusted to enable no entry from outside. In hotels baby sitters may be arranged by the housekeeper and may be members of the staff or from an agency and both should have a written permit from the housekeeper authorizing them to be 'on the floors'; those from an agency will normally collect the permit from the hall porter as they come into the hotel.

Inventories, stock lists etc kept by the housekeeper should help in discovering the loss of items, eg linen, cleaning equipment, etc, through pilfering and as a result investigations should take place.

The housekeeper and her staff should co-operate fully with the security officer over house security regulations. All should realize the need to refrain from gossiping to outside friends and from giving information regarding internal matters to such persons as enquiry agents, newspaper reporters etc.

For security reasons the *housekeeper* selects her staff carefully and prospective new members of staff should be asked for the names and addresses of one or two persons to whom reference can be made, and testimonials should not be relied on. In taking up references, it is wise, if possible, to talk on the telephone rather than to expect former employers to commit themselves on paper.

Health and safety

Fire and personal injury

These are hazards in any establishment and their prevention is of tremendous importance. While the management is ultimately responsible for the prevention of accidents, the *housekeeper*, along with other department heads, should endeavour to see that her staff are safety conscious.

Accidents are costly: there may be serious effects on the injured person; time and materials may be lost; a new employee may need to be trained. Since 1969 employers have been responsible if defective equipment, due to its design or manufacture, causes an accident; compulsory insurance against this came into force in 1970 and there may be other insurance and legal costs.

Poor housekeeping accounts for many accidents and also many accidents occur in an establishment's accommodation area (in one survey taken in a group of hotels it was found there were more days lost due to accidents in the housekeeping department than in any other department). Therefore the *housekeeper* has a great responsibility for making sure that her staff are aware of the common causes of accidents and of the necessary precautions to be taken to comply with the 1974 Health and Safety At Work etc Act.

Under this Act the employer must provide:

and maintain equipment and provide safe working practices;

for correct storage, handling and transporting of articles and substances with maximum safety;

information, instruction, training and supervision to ensure the health and safety of employees;

safe exit and access to place of work;

a good working environment without risk to health and with adequate facilities (WCs, rest rooms etc);

a written statement of general policy which should be displayed by employers with more than five employees and it must be amended as necessary.

Employees should:

take reasonable care of themselves and other employees and other persons on the premises;

co-operate with their employer concerning health and safety.

There is a great variety of accidents causing personal injury which may befall *guests* and staff and while they are normally caused through someone's carelessness they are less likely to occur in a clean, uncluttered and well maintained department.

The following are some of the more frequent safety hazards and causes of personal injury which may occur in the housekeeping department.

Falls

Because of:

frayed edges and worn patches of carpet;
a missing floor tile or uneven floor;
a missing piece from the nosing of a hard stair;
slippery floors, especially in conjunction with small mats;
spillages not immediately dealt with;
tripping over fallen articles;
trailing flexes from equipment, lamps, television, etc;
cleaning equipment left about, buckets, etc;
faulty step ladders;
stools, boxes, etc used instead of step-ladders;
poor lighting in corridors and on stairs;
a step in an unusual place;
no hand grips on baths;
over-reaching;
ill-fitting or inappropriate footwear.

Cuts and abrasions

Because of:

careless placing of razor blades;
careless disposal of broken glass;
careless opening of tins;

absence of kneeling mats for cleaners;
falling objects;
objects poorly stacked and shelves overladen.

Burns, scalds and asphyxiation

Because of:

careless lighting of gas equipment;
absence of fire guards;
carelessness on the part of smokers;
newspapers, periodicals etc, left too near a fire;
careless positioning of portable heaters;
covering of heaters and lamps with clothing, towels and similar articles;
sun's rays striking a concave (shaving) mirror;
hot water bottles being filled direct from gas or electric hot water heaters;
careless filling of hot water bottles from kettles;
too-hot water from shower sprays;
use of certain plastic materials which produce noxious fumes when they
    catch fire;
fire stop doors being propped open by wedges and other articles;
careless use of electric irons and other electrical equipment;
faulty electrical equipment;
misuse of electricity by overloading;
flexes under rugs and carpets.

Electric shock can cause burns and even death and may be the result of

touching bare live wires;
handling appliances which are not properly earthed and so are 'live';
handling appliances with wet hands.

Lifting injuries

May be caused by:

attempting to lift too heavy a load;
lifting incorrectly.

Accidents

Any accident at work, either to *guest* or staff, should be reported
immediately to management or the Health and Safety Officer. It is a legal
requirement under the Health and Safety at Work etc Act 1974 that a record is
kept of all accidents; this is particularly important because of the Industrial
Injuries Act whereby staff may be entitled to claim compensation. It is usual
for the establishment's own accident report form to be completed as well as
the statutory one.

The accident book and report form should be completed at the time of the
accident or as soon after as possible by the injured person or his supervisor. It
should state:

personal particulars of person injured eg name, address, age, occupation;
date and time of accident;
place of the accident;
injury sustained;
cause and/or description of the accident;
what the person was engaged in at the time;
treatment given and by whom;
names of witnesses.

It should be signed by the supervisor and if possible by the injured person.

Accidents to guests should also be recorded and all staff should be aware of the fact that they should *never* accept liability for an accident. Insurance can be taken out by the establishment for protection against claims made by guests on staff.

In the case of fatal accidents, major injuries and dangerous occurrences listed in the Reporting of Injuries, Diseases and Dangerous Occurrences Regulations 1985, the Environmental Health Officer should be notified as soon as possible, preferably by telephone, and details entered in the accident report book. A written report should be sent to the Environmental Health Officer within seven days. These regulations apply to guests and staff.

Prevention of accidents

Unless precautions are taken accidents may easily occur and the *housekeeper* should therefore see that her staff are made aware of the problems and are instructed in the:

use of correct working methods;
need for tidiness in their work;
need for storing things in their right places;
dangers of floor surfaces being left wet, overpolished, etc;
necessity of reporting surfaces and articles in need of repair or replacement;
advisability of wearing suitable shoes, and clothes which are not constricting;
need for warning signs on wet floors;
need for hazard spotting.

A record of training should be kept and be signed by both trainer and trainee and as well as training her staff to be aware of the causes of personal injury, the housekeeper should herself make the necessary reports to maintenance and *follow up* these reports. She should also see that provision is made for:

hand grips on baths;
non-slip mats in showers;
good lighting on stairs and corridors;
help for maids when jobs are heavy or involve much lifting and stretching;
special marks on clear glass doors to prevent people walking into them.

Prevention of accidents may be helped by analysing the accident record book, which may show:

nature of injury;
description of accident;
cause;
who was injured;
prevention (remedy).

Prevention of fire

As in the prevention of personal injury every precaution possible should be taken against fire.

Staff should be made aware of such dangers as:

smoking in bed, in such unsafe places as bedding and linen stores and in areas where cleaning polishes and rags are kept;
leaving chute doors open;
using electric light bulbs that are too strong in lamps;
not reporting faulty electrical equipment, sockets etc;
not unplugging electrical appliances eg TV;
leaving cameras and magnifying glasses where the sun can catch them.

A record of training (including fire practices) has to be kept (as in the prevention of personal injury) and the record signed by both trainer and trainee.

The housekeeper should make provision for:

sufficient and suitable ashtrays;
suitable waste paper bins;
flame resistant and non-toxic furnishing materials (page 213);
proper storage for cleaning rags, linen, rubbish etc;
low wattage lamp bulbs for children.

It is not always possible to stop fires starting but it should be possible to stop them spreading and endangering life. *Prevention, control* and *escape* are three things which require careful thought when considering the risk of fire and bomb scares. Hotels and similar places have particular problems because guests are often unfamiliar with the layout and spend much of their time in the establishment resting or sleeping.

A fire certificate has been required for those establishments or parts of establishments which come within the Offices, Shops and Railway Premises Act 1963. This Act, however, did not cover the residential part of hotels and other establishments and the Fire Precautions Act 1971 was introduced to rectify this situation. This Act makes provision for adequate means of escape and related fire precautions in places of public entertainment, recreation and instruction, as well as those establishments providing sleeping accommodation for more than six persons (*guests* or staff) or where the sleeping accommodation is above the first floor or below the ground floor.

Before a fire certificate is issued the fire authority must be satisfied with such requirements as:

means of escape and whether they can be safely used, eg unobstructed escape routes, the use of emergency lighting, clear signs to exits and fire stop doors;

fire fighting equipment, specific types in specified areas;
means of giving warning of a fire;
staff training;
fire practices and the appropriate records;
fire detectors (smoke or heat);
instructions to *guests*.

While many of these regulations are more the concern of the manager or maintenance department than the *housekeeper,* she should have a knowledge of the Fire Precautions Act 1971 and co-operate with management wherever possible.

She should see that her staff are fully aware of the procedure in case of fire. New staff should be given a fire instruction sheet and it may be necessary to have this in several languages. The staff should realize the importance of:

keeping all escape routes clear;
closing fire stop doors;
closing chute and lift doors;
reporting faulty springs on doors (self-closing);
reporting exit signs not lit;
reporting suspected faulty fire fighting equipment;
reporting any missing equipment;
reporting any missing 'instructions to guests'.

Fire instructions to the *guest* should be placed in the rooms where they are most likely to be seen and more detailed instructions for the staff may be placed in maids' service rooms and similar places and these too should be in several languages. Fire exit notices on corridors etc, should be illuminated from a source other than the main electricity supply. In some cases fire stop doors are not kept closed but close automatically when the fire alarm is set off.

It is essential in public areas and desirable in others that curtains and similar hangings should be of such material or so treated and maintained that they will not readily catch fire. All cotton, linen and most rayon fabrics can usually be given a flame-resistant finish. Pure wool, glass fibre and modacrylic fibres are inherently flameproof.

Fire emergency

In the event of a fire:

operate nearest fire alarm;
attack fire *if* no personal risk;
close windows, switch off electrical appliances;
close door and report to immediate superior;
carry out instructions, eg rouse guests, make sure rooms empty etc;
report to assembly point for roll call;
do not use lifts.

The standard glass-fronted fire alarm is operated by breaking the glass and this sets off the bells, buzzers etc. They may be connected direct to the fire station; if not, the switch board operator should telephone the fire brigade immediately.

Fire alarms may be automatically started by heat or smoke detectors in ceilings. These may be connected to a sprinkler system.

Staff, during their training sessions, should have been instructed as to their exact duties in an emergency as there may be some variation from department to department and establishment to establishment, eg individual staff duties, to whom to report, where to assemble etc.

Fire fighting equipment

This includes:

Buckets of water – easily used but unless they are checked frequently there may be insufficient water in them at the time of an emergency.
Buckets of sand – useful for smothering small fires and may be used if perfectly dry on electrical fires.
Hose reels – more effective than buckets of water or soda acid (water) extinguisher; can extend up to 36 m.
Extinguishers
Soda acid (water)   red   use for wood, paper, fabrics etc
Powder   blue   use for all risks, flammable liquids and gases
Foam   cream use for flammable liquids, oils, fats, etc
BCF (halon)   green   use for electrical and flammable liquids
CO_2 black   use for electrical and flammable liquids
Fire blankets – used for smothering fires.

First aid

Illness, accidents and other emergencies to guests and staff unfortunately occur from time to time in any establishment, and while the housekeeper may or may not be the official first aider she may become involved. Under the Health and Safety (First Aid) Regulations, July 1982, the employer must provide *sufficient first-aid equipment, facilities and personnel and must inform his/her employees of the first-aid arrangements made.* (This regulation regarding first-aid provisions does not apply to premises forming part of a hospital eg residences.) First-aid personnel should be available at all times and staff should know who they are. While a housekeeper should have a knowledge of first aid it is essential that she be level-headed and able to take command of a situation so that it does not become out of hand and to prevent panic, gossip or consternation spreading throughout the house. In order to stop the spread of disquietening facts or gossip, staff should be asked to co-operate and be discreet with the *guests* regarding unfortunate incidents. Inevitably there will be the maid who is anxious to tell the *guests* of an accident or death which occurred in a certain room and, while it may not worry some *guests* it will others, and in either case it would be better left unsaid.

In the case of illness a doctor is normally on call and the *housekeeper* will contact him when necessary, and after the visit she will ensure that his instructions are carried out; in the case of an emergency, 999 may be dialled and an ambulance called. In large establishments, there may be a resident doctor or qualified nurse in attendance (and always a first-aider), and this relieves the housekeeper of much responsibility.

In all establishments there should be accident report forms to be completed giving details concerning any accident which has occurred (see page 61) and in the case of accidents occurring to the staff the accident book required by the Department of Health must be filled in.

On being told of *guests* or members of staff not being well, the *housekeeper* will visit them and see to their needs, helping them in whatever way she can, by making them comfortable, allaying any anxiety if possible, and when necessary calling the doctor and following his instructions. In case of notifiable diseases, eg smallpox, diphtheria, measles, typhoid fever, scarlet fever, poliomyelitis and whooping cough, the doctor notifies the Medical Officer of Health and if the room needs to be fumigated this will be done with an approved fumigant or by the health authorities.

In order to fumigate a room satisfactorily, windows, ventilators, chimney, keyholes, etc need to be securely blocked and the chemicals used according to instructions. On leaving the room, the door should be adequately sealed, usually with adhesive paper and the room left undisturbed for the required period of time. Later the room should be well ventilated and thoroughly cleaned.

In the case of a death being reported to the *housekeeper*, she tells the manager and a doctor is called immediately. The central heating or air conditioning should be turned off and, to prevent unauthorized persons entering the room, the door is locked until the body is removed. The removal of the body should be done as unobtrusively as possible, and often takes place at night, or some other quiet time when there are few *guests* about. In the case of a suspected suicide, any drinking glass, tablets or vomit must be left for the doctor and/or police as they may be needed as evidence.

The first-aid box

First-aid boxes are required to be kept and made available to all members of staff in certain areas of the establishment under the First Aid Regulations 1982. The boxes must be checked regularly to ensure that they do not contain less than the minimum required by law.

A more comprehensive stock of materials may be kept in the housekeeping department and may include:

waterproof adhesive dressings	safety pins
roller bandages	scissors
triangular bandages	antiseptic cream, eg Savlon
sterilized cotton wool	antiseptic/disinfectant, eg Dettol, TCP
sterilized dressings	painkillers, eg aspirin, paracetamol
clinical thermometer	bicarbonate of soda
pair of tweezers	kaolin
eye bath	calamine lotion
sterilized eyepads	feeding cup
pressure bandage	medicine glass
pen torch	bedpan and urine bottle

The first aider, who may be the housekeeper or one of her assistants, will, of course, only deal with immediate treatment or *first aid*, and will leave special treatment or *second aid* to the doctor.

First-aid remedies

The following are some of the possible emergencies or illnesses which could occur and the treatment and remedies given are *first aid* only.

Shock may be caused through an injury giving rise to pain, through haemorrhage or through mental stimulus, such as bad news; the patient is pale and complains of feeling cold and shivery.

The patient should be laid flat, with all constricting clothing loosened, kept warm by covering with a blanket and given nothing by mouth. (An exception is in the case of shock due to mental stimulus, when hot, sweet tea may be given.)

Fainting may be caused as for shock and the loss of blood from the head gives rise to extreme pallor, beads of perspiration and loss of consciousness.

The patient should be laid flat and be prevented from being 'crowded in', so as to get plenty of air, and then be treated for shock.

A heart attack is due to a clot of blood in the heart and manifests itself by an acute chest pain, breathlessness and feeling faint.

The patient should be propped up or allowed to sit forward on a chair and on no account moved until the doctor or ambulance arrives.

A stroke is associated with high blood pressure and there may or may not be a loss of consciousness, but there is usually some degree of paralysis on one side of the face and body.

The patient should be treated for shock and a doctor called.

Concussion is caused by a blow on the head which may or may not render the patient unconscious. If, on questioning later, there is any sign of loss of memory concerning the accident or the time preceding it, then concussion should be suspected.

The patient should be treated for shock and a doctor called.

Diabetes is a disease of the pancreas which prevents the body from burning or oxidizing sugar. Many diabetics are treated with insulin which has to be carefully balanced with the diet. If insufficient food is eaten to balance the insulin, the patient starts to perspire and becomes irritable and nervous. Most diabetics carry a diabetic card and sugar for such emergencies for, should this condition be allowed to continue, coma will result.

The patient should be given two lumps of sugar, a piece of chocolate or any available sweet food or drink at the first sign of insulin shock, and if there is no response, a doctor or ambulance should be called.

Epileptic fits can be major or minor, and it is not unusual for a major fit to follow a minor one. In major epilepsy the casualty will suddenly lose consciousness and fall to the ground, and will then have a series of convulsions, which may be quite violent.

The place where the casualty has fallen should be cleared of obstacles so that the patient does not hurt himself. If possible, clothing should be carefully loosened and something soft placed under the casualty's head. When the convulsions have ceased (a few minutes), the casualty should be placed in the recovery position and watched over until he is fully recovered. The casualty's doctor should be informed of the attack.

Convulsions are fits occurring in young children and babies during teething, and frequently herald the onset of the infectious diseases. The child holds its breath, becomes rigid and purple in the face.

The patient should be kept warm by covering with a blanket, or being placed into a warm bath, and a doctor called.

Asthma results in the patient finding it difficult to breathe and having a feeling of suffocation.

A chronic asthmatic will have had attacks before and may have had drugs prescribed to take during an attack. The patient should be reassured and, if necessary, a doctor called.

Poisoning may result from swallowing, inhaling or injecting poisonous substances.

In most cases when poison has been taken by mouth, the patient should be made to vomit by swallowing warm water with salt or mustard in it, and prevented from sleeping until a doctor or ambulance arrives. If the poison is known to be a corrosive, then vomiting should be avoided, and the patient should be taken to hospital as soon as possible. If a person is found unconscious and an empty bottle which contained sleeping tablets or other drugs is found, an ambulance should be called at once and the bottle kept as it may be needed as evidence.

Burns are caused by dry heat, hot fat or oil, while a *scald* is caused by moist heat.

There are different degrees of burns and scalds, and for minor ones where the skin is not broken the affected part should be immersed in cold water.

For more serious burns and scalds, the air should be excluded by covering the affected part with a clean, dry dressing, and applying nothing else because of the risk of infection. The patient should be treated for shock and a doctor or ambulance called.

Burns may also be caused by clothes catching fire. The flames should be smothered with a blanket, rug or heavy coat and the patient laid down, treated for shock and, if necessary, a doctor or ambulance called.

Electric shock can be caused by a variety of faults in and the mishandling of electrical equipment and may result in burns, shock and even death.

The current should be switched off and artificial respiration applied if necessary. Any burns should be treated and the patient treated for shock and, if necessary, a doctor or ambulance called.

Cuts and abrasions may be caused in many ways and may vary considerably, not only in the extent of the damage, but also in the risk of infection, eg when they are caused as the result of a rusty tin or grit on the ground.

The wound should be cleaned with warm water and antiseptic, and covered with a clean, dry dressing. If bleeding is profuse pressure should be applied on the wound, if there is no foreign body, eg glass or metal, in it; or on the pressure point nearest the wound, between the wound and the heart. The patient should be treated for shock and, if necessary, taken to hospital, or a doctor or ambulance called.

Nose bleeding may be spontaneous or due to a blow, and it is generally more frightening than dangerous.

The patient should be reassured and the soft part of the nose pinched. The patient should sit with head slightly forward and should breathe through the mouth. It should be suggested that the nose is not blown for some time after the bleeding has stopped or the clot may be broken.

Fractures and sprains are generally caused by a fall giving rise to pain and swelling; sometimes there may be bleeding and the bones may protrude.

In the case of fractures, movement of the broken bones may cause extensive damage so, in most cases, the person should not be moved until the doctor or ambulance arrives. The patient should be treated for shock, any bleeding arrested and if movement is necessary, the fracture should be immobilized. The patient should be taken to hospital or a doctor or ambulance called.

A sprain should be bandaged using a crepe bandage, immersed in cold water and the patient treated for shock. If great pain or swelling should occur the patient should be taken to hospital or a doctor called.

Foreign body in the eye may be grit, glass etc, causing pain.

The injured eye should not be rubbed but bathed with the aid of an eyebath and the nose blown thoroughly. If the object can be seen, it should be possible

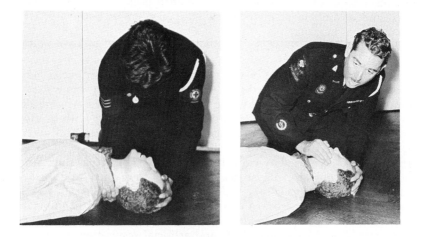

Figure 5.1   Mouth-to-mouth or kiss of life method of artificial respiration

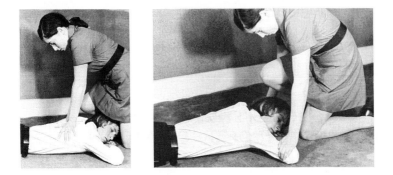

Figure 5.2   Holger-Nielsen method of artificial respiration

to try to remove it with the corner of a clean handkerchief. If not removed and the eye is painful, the patient should be taken to the doctor or hospital.

Artificial respiration

The two most usual methods of artificial ventilation are mouth-to-mouth and the Holger-Nielsen method.

Mouth-to-mouth or kiss of life

This is easiest when the casualty is on his back but it should be started *immediately*, whatever the casualty's position. The casualty's head should be tilted backwards, putting a hand under his neck and pulling the chin upward (see Fig. 5.1), in order to get a clear passageway to the lungs. Pinching the casualty's nostrils hard and forming a seal round his mouth with the lips, blow hard into his lungs until his chest rises to maximum expansion. Watch chest fall and repeat inflation. After two inflations check the pulse to be sure that his heart is beating. If it is, continue to give inflations at 12–16 times per minute until the doctor or ambulance arrives.

Holger–Nielsen

The patient is turned face downwards with his head turned to one side and by kneeling at the patient's head and putting the hands over his shoulder blades (see Fig. 5.2), pressure is exerted by slowly rocking forwards (for an adult the pressure should be about 13.6 kg). As the pressure is released by rocking backwards, the patient's arms are raised by the elbows, to expand his chest, and the process repeated until the doctor or ambulance arrives. Each phase of expansion and compression should last about 2½ seconds, the complete cycle being repeated 12 times per minute.

6

THE HOUSEKEEPER AND THE
MANAGEMENT OF THE DEPARTMENT

If the housekeeping operation is to be efficient considerable thought must be given to the way in which it is organized. One cannot expect a group of people, however well motivated they may be, to give of their best if there is no order or method in the organization.

Organizing is sometimes considered an occasional function, something which is done infrequently, for example when a department is being planned or newly opened or when someone new takes over; but in fact organizing is a continuing function and the organizational 'set-up' should be reviewed frequently.

The *housekeeper* is the man or woman responsible for the efficient and economic running of the department within the aims and objectives as set out by top management. As manager of the department the *housekeeper* has the responsibility of planning and forecasting for the department, organizing, leading, directing, controlling and co-ordinating the accommodation area under her jurisdiction and while doing this must comply with the various legal requirements appertaining to the accommodation department, eg

Fire Precautions Act 1971
Health and Safety at Work etc Act 1974
Food Hygiene Regulations 1970
Hotel Proprietors Act 1956
Sex Discrimination Act 1975
Race Discrimination Act 1976 (rev)
Employment Protection Consolidation Act 1978

The *housekeeper* therefore employs all aspects of managerial activity and, while these may be considered separately, they are very closely interrelated and all are assisted by good communications.

The scope of the *housekeeper's* work varies greatly from place to place and from housekeeper to housekeeper. In the main it is for the organization of the cleaning of the establishment's premises, or such parts as the employing authority dictates (eg kitchens, restaurants and dining rooms are not normally the concern of the hotel housekeeper or the hospital domestic services manager, but they may be of the domestic bursar in hostels), as well as for the management of the staff engaged in the cleaning and servicing of the specified areas. The choice and care of the furnishings also normally come within her scope (see p. 4).

Effective management by the *housekeeper* should lead to:

cleanliness of the premises,
a comfortable and safe environment for the guest,
consideration for the welfare and motivation of the staff,
economic running of the department,
a contribution to the profitability, reputation and smooth running of the establishment.

It is essential for the *housekeeper* to be aware of the aims and objectives of the establishment as a whole and for her to be informed of and consulted on any policy changes which may affect her department. Costs have risen steeply over the last few years and management has to decide what services it can afford to offer and the best way of providing them. Hotels cannot afford to have empty rooms and some cannot afford to offer the services offered in the past, eg early morning teas, 'turning down', or shoe cleaning, nor can university halls of residence afford to have rooms empty during vacations.

A problem which can arise, not so much perhaps in hotels but in other sectors of the industry, is that top management is often indifferent to the real advantages of good housekeeping. While wanting a clean, safe and comfortable environment, top management may be totally unaware of how this may be achieved or of how much it will cost. Objectives and responsibilities are often poorly defined and it is only when housekeeping is an integral part of the whole organization and the *housekeeper* is armed with the necessary information regarding objectives and responsibilities that she is able to set about managing her department efficiently.

Planning and forecasting

First, the *housekeeper* will plan and forecast for her department. She will look ahead and try to predict future happenings, eg staffing for high and low occupancy, annual cleaning, redecoration, etc. She will plan in order that these eventualities are met and that her objectives are reached within the time available. It is possible, of course, that however carefully she plans and forecasts, circumstances arise over which she has no control.

A good planner thinks on the lines of economy, making the best possible use of time, labour and materials and this will be made easier for the *housekeeper* if she has been consulted at the designing and equipping stage of any new or altered building. It should be borne in mind that labour costs account for 90 to 95% of the total cleaning costs and that cleaning and maintenance costs over a period of about twenty years may equal the initial cost of the building, but it should also be remembered that the planning of areas should be as flexible as possible to enable multipurpose use. Designs should be simple, standardized and planned for easy cleaning, as well as allowances made for change eg:

hotel bedrooms cleared for exhibitions, small luncheon parties and meeting rooms;
suites let as individual bedrooms or meeting rooms, etc;

students' rooms suitable for short term 'lets' during vacations when the status and requirements of the *guests* may be different from the students for whom the hostel was first intended;

hospital flatlets suitable for one sister, two staff nurses or four student nurses as circumstances dictate.

In planning and forecasting for the department the *housekeeper* tries to make the fullest and most efficient use of equipment, space and human effort. She plans:

what work has to be done;
when and how often it has to be done;
how it is to be done;
to what standard it is to be done;
how long it will take;
who will do it.

The housekeeper thus concerns herself with staffing requirements and studies the advantages and disadvantages of the whole range of cleaning equipment, agents and methods, and endeavours to bring into use those which make cleaning easier for her staff, save time and are more efficient in producing the final result; in this way she will not only improve working conditions but reduce expenditure and, ultimately, labour costs.

Work study

In setting out to find and implement the most effective use of equipment, space and human effort, the *housekeeper* is making use of method study – this is part of *work study*, a 'tool' of management. Work study also includes work measurement, which is required to determine the work involved in a job; measurement is made of the time taken to carry out a job under normal circumstances by an average worker and this may help in determining the number of staff required, in determining who is over- or under-employed and in standardizing labour costs.

Work study has been applied to various aspects of housekeeping, eg bed-making, the planning of the linen room and its work, general cleaning procedures, etc. In a particular investigation it was shown that the distance covered by the maid during bedmaking could be reduced considerably if she stripped the bed more systematically and did not tuck in the sheets and the blankets until the end of the bedmaking operation.

From other investigations, it seems that block cleaning, rather than a room being completed in one visit, has resulted in better working conditions (less fatigue) and better work flow for the staff as well as a saving of time. There is, however, the question of security to be considered.

Individual areas should be planned with a view to the work that will be carried out in these areas and to their relation to the rest of the establishment, eg maids' service rooms, lifts, linen rooms, etc.

Work study should be considered wherever a wastage of time, labour or materials is suspected, eg when:

delays occur;
equipment lies idle;

work schedules appear unsatisfactory;
overtime appears excessive;
quality of work is poor;
there is a high level of fatigue;
turnover of labour is high;
workers are not fully occupied;
rate of absenteeism or accident is high;
unnecessary movement is suspected;
guests complain of delays, etc.

Sometimes investigations are carried out by trained personnel instead of the *housekeeper*, but before any investigation is started a full explanation of the need for work study and the way in which it is to be carried out should be given to the staff.

It is not necessary to go into details here as to how work study is carried out, but the main steps are as follows:

1 The job procedure is selected and the problem defined.
2 The present method is recorded by the use of
   a) outline and flow process charts
   b) flow and string diagrams.
3 The findings are examined.
4 The improved method is developed.
5 The improved method is installed.
6 Periodic checks are made to ensure the 'improved' method is working satisfactorily.

In any job the best results are obtained after practice. Once the new time and labour saving methods have been accepted the gain will no doubt become, apparent, but the change from the 'customary' methods often takes time and great tact. The staff should be made fully aware of the new working methods and the reasons for the change. Only with the full co-operation of the staff throughout the investigation is it possible to get a complete picture of the old working methods and to install the improved methods satisfactorily.

Standards of housekeeping

Methods used and the time taken on any job will inevitably affect cleaning standards and the *housekeeper* has to plan a standard of cleanliness. It is not always necessary to have the same standard throughout an establishment and this is most clearly illustrated in hospitals where there are areas of high, medium and low risk (see table overleaf).

A standard of cleanliness is almost impossible to measure and while such measurements as dust or bacterial counts can be made they are more suited to specialized purposes, eg hospital operating theatres, than for general use.

Other measurements have been based on the number of square metres cleaned per worker, on the annual costs per square metre or other circumstances. These measurements based on statistics, without a full knowledge of the facts on which they are calculated, can be very misleading. Very few areas are identical in:

degree and type of soiling,

Standards of cleaning in hospitals

Area	Standard	Requirement
high risk areas, eg operating theatres special units etc	prestige standard	highest possible standard of cleaning, appearance, dust and infection control
medium risk areas, eg wards, sluices, toilets, kitchens	special standard	high standard of cleaning, appearance and infection control
low risk areas, eg corridors, offices, residences	normal standard	good standard of cleaning and appearance, absence of soil

amount and type of furniture,
furnishings,
obstructions, etc.

But these are only a few of the variables and so the figures can only be 'average'. As a result faulty and expensive decisions may be made. More often than not, no measurement of cleanliness is taken but the quality of cleanliness is based on the acceptability or unacceptability of the work.

The most practical definition of housekeeping standards is:

method × frequency

Therefore, an acceptable standard should be obtained when:

cleaning methods are correctly selected;
correct equipment and agents are used for each surface involved;
cleaning tasks are carried out at frequencies dependent on the type and amount of soiling, which may detract from the appearance of an area and may put the occupants at risk of infection.

These points should be included in the final documentation following method study.

Perhaps more than in any other type of establishment it is of vital importance in hospitals that there is full consultation when establishing methods and frequencies. In patient care areas the views of the nursing staff will carry considerable weight, but the responsibility for determining the ways in which the standards should be met rests ultimately with the domestic services department.

The *housekeeper* should therefore plan for a standard at the level desired by management, making use of her technical knowledge in defining job procedures, job sequences and frequencies, and work schedules for her particular establishment. The most usual method of ensuring that standards are being met is by establishing a system for checking work done (see pages 45–9). Inspections may be total, random or planned and the housekeeper may also introduce quality control when check-lists and 'white ragging' can be used in an attempt to compare the work with an ideal standard. The result may be judged as a percentage of the ideal or as fair, good or excellent. Effective systems of checking are essential to ensure that all work is carried out and that standards are maintained.

The housekeeper may need to reassure her staff that checking the quality of work is a means of improving performance and not an excuse for fault finding. Staff meetings may provide the housekeeper with opportunities to discuss standards with her supervisory staff and some means of feedback to those carrying out the tasks, eg maids, cleaners, etc, should be found. The *housekeeper* should find time to carry out inspections herself as part of her monitoring function but generally it is work delegated to the assistant housekeepers or supervisors.

The way in which a job is done (job procedure and sequence) affects the time spent on that job, as well as cleaning standards, and cleaning time is the basis from which staffing requirements stem.

Standard time rates have been calculated for specific jobs under 'standard' conditions of equipment, agents, method and personnel etc, but these seldom exist. The rates published cannot take into account all the factors which influence the time required for a particular job in a particular area in a particular establishment. Standard time rates, however, may be used as guidelines or for comparative purposes. Among the factors which will influence the 'standard' time needed for any job are:

the type, age, architectural features of the establishment;
the function of the area;
the maintenance of the area;
the standards to be obtained –
  degree and type of soiling,
  frequency of cleaning,
  type of surface to be cleaned,
  type of service to be rendered;
the amount of traffic and interruptions;
the habits of the occupants;
the accessibility of work areas to service areas;
the availability and type of equipment, supplies, etc;
the dexterity, motivation and calibre of the employees;
the quality of supervision.

Staffing requirements

Taking into account the particular circumstances, staffing requirements can be adequately calculated when there is a sound knowledge of average 'standard' time rates needed for any job procedure or sequence. This may be done by totalling all the times for the jobs to give the gross man-hours (or minutes) per week (or day). This figure divided by the number of hours (or minutes) to be worked by each member of staff will give the total number of staff required.

Suitable allowances should be made for mid morning or other breaks, preparation and cleaning of equipment etc, if not built into the 'standard' time. An allowance of 50–60 minutes in a day would not be unusual.

So, assuming daily work takes 4470 mins.
and the cleaners work 8 hours a day – 480 mins.
and the allowance for breaks, etc is – 50 mins.

the actual working time per cleaner is (480–50) – 430 mins.
then, the number of full-time staff required is

$$\frac{4470}{430} = 10.4$$

ie 10½ full-time employees (FTE) would be required to cover that work.
10% is normally allowed for holidays, sickness, etc,
so, number of staff = 11½ FTE working 8 hours a day
or 23 four-hourly PTE (part-time employees)

When these calculations include time for daily work only, then extra staff will be required to cover periodic cleaning. The calculations are often based on annual hours and in many cases there will be one cleaner required for periodic cleaning to every five for daily work. Thus in the example above 2½ FT cleaners would be required for periodic cleaning. However the standard and frequency of periodic cleaning vary greatly and will affect the number of staff required.

There are situations where it may be more beneficial for some work to be done on a cleaning contract rather than by direct labour and the *housekeeper* will have to plan the amount of direct and contract labour required for an economically run department (contract cleaning, see p. 50).

Where all floors of an establishment are identical, and the standard of work throughout is the same per area and per resident or guest, it is not difficult to develop work loads once the staffing requirements and the hours the staff are to work, ie whether FT or PT, have been established. For example:

In students' hostels or hospital residences the cleaners may each be responsible for the cleaning of 20 study bedrooms, washroom block, utility (amenity) room and corridor in a FT duty period, but if there are different grades of residents and varying types of accommodation the allocation of work is not so simple if the work load is to be shared equally. (If shorter hours are worked each cleaner will be responsible for a smaller section but more cleaners often mean more money, eg overalls etc.)

In some places the number of hours per week for the different types of work may be calculated and work allocated accordingly:

eg for rooms requiring
   bedmaking and service – 3–3½ hrs weekly
for rooms requiring
   bedmaking but no personal service – 2 hrs weekly
for rooms requiring
   no bedmaking or personal service – 1½ hrs weekly
for vacant rooms (kept ready
   for occupation) – ¼ hr weekly
for bathrooms – ½–¾ hr weekly

In hotels a room maid may have a section of 12 bedrooms with private bathrooms but problems arise due to: a) daily changes in room state, ie in the numbers of departure, occupied and vacant rooms; b) staff shortages due to sickness and days off; c) economic restraint on the wages bill. So the allocation of work is difficult and flexibility of staff is essential.

In these cases, a points system may be instituted by which each type of accommodation is given a different point rating (eg single room 4 points, double room 5 points, twin bedded room 6 points) and each maid given the same work-load according to the point rating of the rooms she services. In other cases a maid may be given an extra room to service for every vacant room in her section.

Area/job assignment

In most cleaning operations the *housekeeper* arranges the work on an area assignment basis. The individual maid or cleaner is responsible for one area or section of rooms. Area assignment may lead to:

greater pride in the work;
a competitive spirit between those working in similar areas;
better security;
easier supervision as the maid is more easily located.

There are, however, some cleaning operations planned on a job assignment basis where staff may be working individually or as a team, on a single job throughout the establishment or over a wide area. Such jobs may be carpet shampooing, window cleaning, paint washing and floor cleaning. With this system there may be:

specialization and so greater efficiency;
a saving in training and equipment;
more difficulties with security and supervision.

It cannot be over-emphasized that the standards of work which can be achieved in a given time, by a given number of employees, will vary greatly from place to place. However, too often, insufficient thought and planning is given to staffing requirements (staff establishments) and as a result the best is not always received from the staff employed. Work has a tendency to 'expand according to the time available for it' and where the *housekeeper* introduces more easily cleaned surfaces, newer equipment etc as they become available, it is possible, without reassessment of the situation, to become overstaffed or for there to be a poor allocation of work load, with consequent discontent amongst the staff. The *housekeeper* must use all the 'tools' of management to deploy staff economically and efficiently.

Reasons for the uneconomic running of her department may include:

wastage of labour – too many staff;
– outdated jobs;
– outdated equipment;
– little mechanical equipment;
– outdated materials;
– outdated designs;
– stock not being used (money lying idle);
– lack of supervision;
– lack of planning.
wastage of materials – extravagant use of cleaning agents;
– insufficient cleaning of articles;

- insufficient care of articles;
- incorrect methods of cleaning;
- extravagant use of heat, light and water.

Organization

In putting her plans into operation the *housekeeper organizes* the work of her department, when personnel and staffing problems take up a great deal of her time. The staffing of the housekeeping department involves recruitment, selection, training and welfare of the staff, as well as embracing other aspects of staff management such as the skills of delegating work and of communicating with, motivating and supervising the staff.

Recruitment of staff

Even when there is a personnel officer or staff manager, the *housekeeper* has the final decision regarding the selection of her staff. Where she is responsible for recruitment, appointments should be made as quickly as possible other-wise existing staff may become uncooperative about the extra work to be done. The following may be useful sources:

1 Advertisements in newspapers and trade journals – these should be as detailed as money allows and if a box number is used, the locality should be indicated. The most suitable papers and days may become known and advertisements should not be kept in too long or they may lose their effectiveness.
2 Private agencies – this may be an expensive method of recruitment as the agency normally demands a high fee from the employer.
3 Local Job Centre. It is useful to keep in contact with these people so that they know the *housekeeper's* particular requirements.
4 Cards in shop windows – this is a useful method of recruitment for local part-time labour.
5 Internal grapevine – this may lead to the employment of friends and relatives of existing employees, which can have advantages and disadvantages.
6 Former employees – the majority of workers leave voluntarily and over a period of time home ties and other circumstances change.
7 Colleges, schools, etc – courses normally finish about the end of June, but there may be students available for evening, weekend or holiday jobs.

It may be useful to record the result of the methods employed for future reference and the headings seen in the table on page 79 might be appropriate.

Where possible, any suitable applicant should be called for interview and the following points are suggested for an inexperienced interviewer:

1 Full name and address – if the applicant is to be non-resident, travel difficulties may need to be pointed out.
2 Age and place of birth – these should indicate whether the applicant is too near retiring age, too young for the rest of the staff, needs a work permit.

Source of Recruitment	Cost	Date	Response	Applicants employed

3 Qualifications and particulars of former jobs; reasons for changing.
4 Health and sick leave in the past – such things as shortness of breath, varicose veins etc, may be noticed and lead to further questioning.
5 Family commitments – these are of importance where the applicant is to be non-resident and cover such points as partner's work, age of children.
6 Work and other relevant details discussed, such as – hours, duties, wages, holidays, uniform, meal facilities, any deductions to be made from pay, eg sick scheme or superannuation, regulations concerning body searches, the searching of employees' bags as they leave the premises, length of time for notice of leaving or dismissal (it is usual for a week's notice to be given on either side) and, if to be resident, the type of accommodation should be shown where possible.
7 The applicant should be given an opportunity to ask questions.
8 Names and addresses and/or telephone number of referees.

During the interview the *housekeeper* will form certain impressions about the applicant, eg neatness, health, personal hygiene, politeness, cheerful or moody disposition and from this knowledge the *housekeeper* should be better able to decide where and with whom the applicant could work.

Selection of all staff is of major importance and the success (or failure) of the selection will be shown in the attitude of the workers. For those at supervisory level, eg assistant housekeepers, it is not always easy to assess their supervisory qualities until they are on the job and poor selection may result in dissatisfaction in those for whom they are responsible leading to:

excessive grievances;
unpunctuality;
absenteeism;
high turnover of staff;
lack of interest;
non-co-operation;
poor standards.

After the *housekeeper* has summed up the suitability of the applicant for the work, and having asked for the name, address and telephone number of a former employer and/or another referee, she will take up references. Many employers are often unwilling to commit themselves on paper, particularly in the case of doubtful employees, so it is not unusual to obtain a verbal reference by telephone. However, this would not provide a permanent record, so the referee should be asked to follow up the conversation with a letter which can then be filed. If the references are taken up by telephone towards the end of the interview, it may be possible to engage the applicant straight away. She should be told the date she is to begin work and asked to bring from her last employer the tax form (P45) showing the amount she has earned and the income tax paid. It is advisable for a *housekeeper* to have a knowledge of the legal aspects regarding such things as the employment of foreign labour, National Insurance, contracts of employment and union requirements.

Job descriptions

Recruitment and interviewing of staff may be easier when a *job analysis* has been carried out and a *job description* prepared beforehand. In job analysis the job is studied in all aspects, including specific tasks performed and its relation to other jobs and work conditions. Job descriptions give a general description of the work to be carried out and of the type of person required to do it, eg educational background, work experience etc. The table on page 81 shows a specimen job description for a floor housekeeper.

Having selected her staff, the *housekeeper* should:

give them the best working conditions possible;
see that they have the right 'tools' for the job;
train them for the job;
motivate them to the job;
supervise them on the job.

Training

Training should start with an induction period which will include:

things explained, eg conditions of service, pay etc,
the need for personal hygiene, courtesy, security, safety and fire
   precautions,
observations to be made and reported (see p. 31)
places shown eg linen room, stores, various offices etc,
people met, eg supervisors.

Even the experienced maid or assistant housekeeper changing jobs can benefit from much of this induction period; layout of the building, people and house customs may all be different from her previous job. Following the general introduction, if the induction period is long enough the maids may be given some off-the-job training in technical skills by a trained instructor. With all off-the-job training it is as well to remember the old saying

I hear – I forget
I see – I remember
I do   – I understand

and so talks should be kept short, demonstrations given and ample time allowed for employees to make full use of cleaning equipment and cleaning agents, methods for the different jobs and to ask questions.

Whether there has been any training in technical skills during the induction period or not, on-the-job training will follow. This may be carried out by 'working with Nellie' but this is hardly an adequate method unless Nellie is a

Specimen job description for a floor housekeeper

Job title	Floor Housekeeper
Place of work	Woodlands Hotel, Housekeeping Department
Purpose of job	1 To supervise Room maids' and Houseporters' work on an allocated number of floors and to ensure that work is carried out to the standard required by the Executive Housekeeper 2 To ensure that departure rooms are serviced and made ready as soon as possible in order that reception may re-let at any time 3 To anticipate the guests' requirements at all times thereby ensuring comfort and satisfaction 4 To provide information essential to management
Hours of work	7.30 am–4.30 pm ⎫ operated on a rota system 5 days 8.30 am–5.30 pm ⎬ per week 2 pm–10 pm ⎭
Responsible to	Executive Housekeeper
Responsible for	Room maids, Houseporters, Cleaners, Cloakroom Attendants
Liaison with	Receptionists, Head Floor Waiter, Valets, Linen Room Staff, Hall Porter and Storekeeper
Scope of Work	Help with training of staff Maintaining stocks Maintaining set standards of work Planning of work Reporting to Executive Housekeeper, or her deputy, problems in carrying out the job
Routine duties	1 Checking staff on duty 2 Issuing keys 3 Supervising room maids, houseporters and cleaners 4 Checking all rooms on floors 5 Issuing cleaning stores 6 Supervising linen requirements and checking floor stocks are correct 7 Keeping in constant contact with reception 8 Reporting maintenance work 9 Attending to guests' requirements 10 Reporting immediately to the management and/or security any persons who may be acting suspiciously and any other security or safety hazard 11 Undertaking any job delegated by the Executive Housekeeper or her deputy
Occasional duties	1 Supervising the changing of curtains, bedding etc 2 Attending staff meetings 3 Further training of staff 4 Helping with stocktaking 5 Planning of extra work 6 Attending fire drills

willing and carefully selected worker who might be given an incentive bonus. An alternative method would be by the use of 'order of work' cards or a training manual under the direct supervision of an assistant housekeeper; better still would be 'on-the-job' training by a trained member of staff, possibly a training housekeeper.

Further off-the-job or refresher training should be included in any training programme and becomes necessary when:

standards are not being met,
there is an increase in costs,
there are changes in policy, house custom, equipment, etc

In all cases of training a record should be kept and signed by the employee and the trainer.

The Health and Safety at Work etc Act 1974 and the Fire Precautions Act 1971 both put a duty on employers to train staff and on employees to co-operate with their employers. Whatever safety training is undertaken it should form an integral part of the employee's training programme and should be included in the induction, on-the-job and refresher training. Once a working method has been determined regarding the use of safety equipment it should be:

taught to newcomers on arrival;
passed on to experienced staff at refresher training;
included in company standards.

The training of staff should be of benefit to them, the guest and management. The reasons for training are to:

enable staff to settle in more quickly and feel that they belong;
show staff the importance of their job and what is expected of them;
show staff consideration is being given to their conditions of work;
maintain standards for the satisfaction of the guest;
prevent the waste of time, labour and materials;
prevent staff fatigue by use of incorrect methods;
increase the skills of the staff;
prevent accidents;
increase the sense of teamwork;
improve supervision.

In some spheres the Hotel and Catering Training Board is trying to improve training and redistribute money for this purpose throughout the country. Its objectives are outlined by the Industrial Training Act. The National Health Service is outside the scope of this Act and the National Health Service Training Authority draws up training strategy and offers advice.

Procedures and schedules

Job procedures specify the way in which a job is to be done (order of work cards p. 21) and should help in training staff in correct working methods. They should be planned in relation to a specific establishment, taking into account different surfaces, different cleaning equipment and agents and they should help in deciding the number of staff required. In addition they should be planned to:

aid standardization;
preserve surfaces and materials;

effect a saving of cleaning equipment and agents;
effect a saving of employees' time and energy;
prevent accidents;
help training;
ensure the completion of a job;
aid the compiling of work schedules and help in staffing requirements.

Work schedules list the actual work to be undertaken by particular members of staff during a particular period of the day. Times of meals, breaks and any special jobs are pinpointed throughout the period so that there is a guide not only as to what has to be done but also when it has to be done. Order-of-work cards may supplement work schedules. Work schedules should be essentially simple in form with the minimum number of words and should be planned to:

make the best use of staff;
ensure coverage of work;
ensure fair allocation of work.

The following is a specimen work schedule for a Domestic Assistant in a hospital ward.

7.05 am	Dust and vacuum sister's office
	Vacuum wards and clean treatment room
	Collect breakfast trays, water jugs and glasses
	Damp dust wards; replace locker bags. Return water jugs and glasses
	Prepare trolley for coffee
	Serve coffee
10.15	Coffee break
10.45	Collect coffee cups
	Clean bathroom, sluices and lavatories
	Thoroughly wash hands
	Lay table and trays for lunch
	Prepare kitchen for service
	Serve after lunch teas
	Clean kitchen
1.30 pm	Lunch
2.00	Extra work on regular days:
	wash bedtables
	wash wheelchairs
	wash flower vases
	polish furniture
	tidy and clean linen cupboard
	change curtains
	Prepare and serve teas
	Collect trays
	Wash out cloths, put equipment away
3.55	Off duty

Duty rosters

In order to enable the right member of staff to be on duty at the right time *duty rosters* are compiled. These should be clearly laid out, showing the hours of duty and days off for each member of staff as well as any other relevant details such as mealtimes. They should be planned to:

enable areas to be covered for the correct periods;
see whether there is over or under-staffing at any particular time;
ensure fair allocation of hours and days off.

Therefore when compiling a duty roster the *housekeeper* should consider the following points:

the type of establishment, guests' requirements, services to be rendered, coverage to be required;
statutory regulations, eg maximum hours, spreadover, meal breaks, days off, etc;
split duties avoided where possible;
maximum staff at peak periods, adequate at other times;
the possible need to rotate the roster;
even share of duties, possible relief for days off, holidays and sickness;
time for training;
due warning given of changes of roster.

In these ways the efficient running of the establishment and the welfare of her staff can be borne in mind by the *housekeeper*. An example of a duty roster is given on page 32. To make an even share of duties the roster should rotate every five weeks. A maid is entitled to be paid overtime if she works more than the statutory maximum number of hours, eg 40 hours a week depending on wages paid, and may also be entitled to extra money for spreadover, ie when work spreads over more than 12 hours.

Welfare of staff

As a manager, the *housekeeper leads* and *directs*; she gives instructions, trains and motivates her staff to meet the required standards. Incentive bonuses may be useful short term motivators but the *housekeeper* should look further than this. Good staff relations and good working conditions are probably longer term motivators. Due consideration should be given to wages, holidays and distribution of hours on duty. Maids should be compensated in some way for extra work done, as for example when there are no relief maids for days off, sickness or holidays. It might be suggested that to prevent monotony maids change sections occasionally, but in the majority of cases maids object because they get the feeling work is left for them and they prefer to keep to their own supervisor and guests.

The Catering Wages Order, the Whitley Council and other bodies lay down minimum wages and holidays, and maximum working hours per week for different types of establishments and different grades of workers; and the housekeeper should be aware of these regulations and comply with them when planning duty rosters and holiday lists. The planning of holiday lists can be a real problem as a minimum number of staff is required at any one time, but maids may have husbands at work, children at school and friends with whom they wish to spend their holidays, so holiday arrangements should be made early with as much consideration given to the individual as possible.

The feeding of her staff is not really the province of the housekeeper but in the interests of their welfare she is concerned that they have sufficient time to cover the distance between working areas and staff canteen, time to queue for, and to eat their meals.

Good employees are hard to come by and once found it is up to the housekeeper to be concerned with their welfare in order that they will stay.

Living-in staff need comfort and warmth, single rooms if possible, as well as a lockable cupboard, facilities for laundry and for making a hot drink, to encourage them to stay. For security reasons it is necessary to have individual lockers in which non-resident staff may keep their overalls, outdoor clothes and handbags.

Unless a *housekeeper* recognizes each member of staff as an individual and not just as another pair of hands there may be instability and discontent amongst the staff. This may give rise to resentment showing itself in the breaking of rules, absenteeism or stirring up trouble. In order that staff are made to feel that they and their work are important the *housekeeper* should show an interest, and offer praise where it is justified. In this way a happier atmosphere is created and less ill-feeling results when it is necessary to find fault. Any fault-finding should be done in private.

In addition to knowing the names of her staff, the *housekeeper* should know something of their lives apart from their work and this knowledge will enable her to understand, sympathize and make allowances in individual performance, when and if necessary.

In planning well the *housekeeper* improves staff relationships and should at all times act as an example to her staff and provide them with the cohesive force of leadership and purpose. She should ensure discipline is kept at a reasonable level, consistent with managerial policy, and make her staff aware of the need to take their share in the efficient and economic running of the department. She should not make more rules for her staff than is absolutely necessary, but those that she makes must be enforced.

Staff have more faith in the *housekeeper* who shows command of the situation and this applies not only to the usual work of organizing the department but also to the way in which she deals with emergencies occurring from time to time. These emergencies could be, for example, fire, accident, death or birth, and in all cases the *housekeeper* is expected to keep a cool head and to maintain discipline and control over staff and *guests* according to house policy.

Co-ordination and control

To ensure that everything works to give a balanced, effective organization the housekeeper needs to *co-ordinate* the activities of the department. She should keep the department running smoothly, dealing with problems and queries as they arise, giving consideration to *guest* and staff welfare and maintaining liaison with other heads of departments. Effective means of communication are of vital importance.

In *controlling* her department, the *housekeeper* constantly checks performance and work results. This involves keeping an ever watchful eye on the work in progress and the costs incurred and collecting information regarding the work from her assistants. Housekeeping is a difficult field in which to exercise control because employees work individually and the 'end-product' is intangible. Planning and control are complementary and where the *housekeeper* finds any deviation from the original plan she should take the necessary steps to remedy it or she may consider it advisable to replan.

In aiming at an efficiently run department with operating costs as low as possible the *housekeeper* endeavours to save time, labour and materials. In doing so, she controls:

work methods;
allocation of work;
working conditions of her staff;
all articles in use within the department;
   eg linen and uniforms,
   keys,
   furniture and furnishings,
   equipment and supplies.

and ensures as far as is possible throughout the department:

the prevention of accidents (page 61)
the provision of first aid (page 64)
security (page 54)
prevention of damage by pests (page 145)

The *housekeeper* carries the direct responsibility for achieving the aims of her department and the only way she can have effective control is by close and careful supervision. While she delegates much of the routine work and day-to-day supervision of the department to her assistants, she should spot check some rooms from time to time and remain observant and perceptive and be, at all times, someone to whom the staff can look up and turn for advice. Leadership is an important element of control and while control is often considered last it is by no means the last step in management or supervision. Control has the greatest impact when applied at every step of housekeeping organization.

(*Note* In smaller establishments the *housekeeper* is much more concerned with the day-to-day routine work and at times may have no assistants on duty with her.)

In surveying and controlling the work of her department, the *housekeeper* should keep abreast of new products, furnishings, uniforms, etc and in doing so she should try out new materials, equipment, supplies and the like in an endeavour to keep costs as low as possible. New ideas may be gained by visiting exhibitions and other establishments, by reading trade magazines and by seeing representatives from various firms when they call. These are all time consuming but essential if up-to-date products and methods, etc are to be used and operating costs kept down. Control of departmental costs is essential and for this to be effective the housekeeper is concerned in the preparation of a budget for her department (see p. 90).

Record-keeping

Paper work is necessary and, although time consuming, the *housekeeper* will need to keep certain records in order to aid memory, to aid co-operation between departments, to improve efficiency and to make it easier should someone have to take over her job. While some records are legally required others may be company policy. Not all records are relevant to all types of establishments, but amongst those kept may be:

records of recruitment for staff and their results;

records of staff, giving personal particulars, eg date of commencement of employment, next of kin, holidays, sickness, absences, date of leaving with reasons and possibly brief notes on their work and conduct in case of requests for references;

record of hours worked by staff, including overtime;

record of staff training;

stock books for linen and stores;

inventories of rooms and equipment, with dates of receipt, cost and possibly a record of maintenance of the individual items;

records of each room regarding redecoration, new furnishings and annual cleaning;

blanket book;

lost property book;

record of missing articles and articles not immediately found, eg vacuum cleaners, etc;

accident book;

record of fire practices;

record of articles moved from or to rooms;

record of individual personal tastes of frequent guests and VIP's;

financial records, invoice and petty cash books, costs of personnel, room servicing, cleaning, contracts, purchasing of equipment and supplies, etc.

Communications

In any establishment there are times when staff and *guests* need to be contacted and unless there is a good system of communication tempers may become frayed, and time and energy wasted. Communications may be required between *guest* and staff, staff and staff and *guest* and *guest*, when the giving and receiving of information, instructions and complaints as well as the dealing with requests and emergencies may take place.

The telephone system is probably the most frequently used means of communication and apart from 'memos' the only form for *guest* to *guest*. The larger the number of instruments strategically placed the more effective but the more costly the system will be. (There are hotels where telephones are installed in private bathrooms and even some where there are two instruments provided in double bedrooms.) Contact may be made by direct dialling or through the switchboard operator. Where dialling is direct the calls may be internal or external; in the latter case in hotels the charges may be 'clocked' up automatically to the occupant of the room. (It would appear that in this case there is no measure of control over the misuse of the telephone by unauthorized persons. The meter is normally only read on the departure of the guest.)

Telephones can lead to many interruptions and where messages are being received continuously it may be convenient to have them recorded on tape, when they may be played back as required. In hotels messages between the housekeeper and the receptionist, the guest and the valet, etc may be recorded in this way. An electro-writer may also be used between one department and another when the machine transmits and records handwritten messages.

In hotels different coloured lights or a combination of coloured lights indicate the state of the rooms on a room status board to the receptionist and the housekeeper, ie whether the rooms are vacated, being serviced or ready. When a guest has paid his bill the light is operated by the receptionist or cashier, and first the maid and then the housekeeper override this light with a 'jack' placed in a socket in the guest's room. In new large hotels this system is computerized; there is then only one light by the side of each room number on the board and by pressing the appropriate button, eg rooms being serviced, the receptionist or housekeeper can obtain the necessary information regarding the rooms. There may be VDUs (visual display units) in such places as the housekeeper's office giving up-to-date status of the rooms. It is possible to recall the information when necessary and at certain times 'print-outs' follow. In future, the front-of-house computer may be linked to each room using the TV cable network and this could include early calls, room status, message-waiting lights, room checking and the activation of air conditioning, smoke detection etc, some of which are already computerized. A small light by the telephone in the room may indicate to the guest that a message is waiting at reception.

Two-way communication is possible in some places where a master console unit is to be found in some convenient place – Sister's office in a nursing home or hospital, reception in an hotel – and a small cabinet is in the patient's or guest's room. In some hotels there is an intercom-cum-baby listener and for the hard of hearing a flashing light fire alarm and, in the event of night-time emergencies, a vibrating mechanism installed in the bed to wake the guest up.

Press-buttons in rooms or by beds in hospitals and nursing homes operate lights or bells in convenient places (eg corridors, service rooms, etc), indicating that a particular service is required. At one time there was always a bellpush in hotel bathrooms, but in many new buildings this is no longer the case. Except where there is a telephone within easy reach this would appear to be a retrograde step as accidents and emergencies do occur in bathrooms. A bell system is, of course, used to raise the alarm in the case of fire or a breakdown in lifts and fire alarms may be connected direct to the fire station.

Paging systems are other means of communication and only in luxury hotels are there page boys or messengers. Public address systems are still used in some establishments but these can prove disturbing. The 'bleep' is a limited paging system; in this case certain members of staff whose work takes them all over the establishment (eg the *housekeeper*) carry a small battery operated device which 'bleeps' when the particular person is required. It is worked via the switchboard and the member of staff 'bleeped' goes to a telephone to find out why he or she is wanted.

Memoranda are means of leaving messages for a person, or of confirming some verbal communication. The visual communication is often better than the verbal one and in hotels there are many, eg arrival and departure lists, room occupancy lists, slept-outs, early morning tea lists, etc. These communications are between department and department, but in addition there are communications found in hotel bedrooms (including fire instructions) designed to be informative in order that the guest may use the services of the hotel to advantage. A folder containing such communications as 'Do not disturb', 'Please clean my room', 'Room service order', 'Directory of services', 'Laundry and valeting service and price list' can prove extremely useful to the guest and save the staff a great deal of time. Some of these communications may be in several languages.

While communications are not a 'tool' of management, good communications are essential for the smooth running of all establishments. Therefore any person in a managerial position, such as the *housekeeper*, is as concerned with communications as she is with the engagement and welfare of staff, the planning, organization, budgeting and the economic running of the department.

7

BUDGETING AND BUYING

Budgets

A *budget* is a plan of expenditure and if there is to be any control of costs throughout the establishment, budgeting is essential. The *housekeeper* is one of those concerned with the preparation of a budget for the department.

The housekeeper estimates the expenditure for the department for a specified period, which is generally a year. The longer the financial period, obviously, the more difficult it is to forecast the requirements and relevant costs for the department.

There are three broad areas to be considered when preparing the budget:

- wages and salaries;
- operating costs (supplies, services, etc);
- capital expenditure (equipment usually with a life of five years or more or over a fixed sum of money).

The *housekeeper* should consider past records regarding occupancy rates, wages and salaries, purchase of equipment, furnishings etc, in an endeavour to forecast future labour requirements, increases in wages, new services being planned, replacement of equipment, linen, furniture, furnishings, etc.

The budget should therefore include the probable cost of:

wages, salaries and contracts;
additional cleaning equipment, furniture, furnishings etc;
replacement of cleaning equipment and agents, guests' supplies, furniture, furnishings, linen etc;
repairs;
in some cases, planned renovation and redecoration.

A priority rating should be made for each item and an indication as to whether an item is a replacement, an improvement or an addition. Higher priority ratings are normally given to replacements than to additions.

Management requires to be furnished with records, observations and evaluations of past performance and plans for the future, so the *housekeeper* should be preparing throughout the financial period for the presentation of the next budget by keeping records of relevant facts, eg the usage of various agents, the cost of repair services etc. In this way she is more easily able to justify the cost of particular items.

The *housekeeper* should realize that changes may occur in such external factors as the labour market, the commodity market and in legislation, and these may all have a bearing on her estimation of the expenditure of the department over the next financial year.

The *housekeeper* should be willing to offer advice (tactfully where not requested) on such matters as the choice of floors and other finishes, furnishings, furniture and fittings to ensure that due consideration is given to economy in the work of the department.

An advantage in preparing a budget is that it provides the opportunity to take a critical look at the costs of the department, review past planning and present accomplishments and then to take appropriate steps to accomplish more in the coming financial period. In this connection it may be beneficial for the *housekeeper* to look more closely at the cost of such things as:

servicing of a room;
cleaning of a particular area;
serving of early morning teas;
night service, ie 'turning down';
overtime compared with extra staff;
hiring compared with owning linen, equipment etc;
checking of linen;
using non-iron linen with or without laundry on premises;
office supplies, handwritten versus duplicated versus printed lists;
re-upholstering versus purchasing new;
use of contracts;
bulk buying.

Buying

In any establishment there should be a policy in regard to depreciation and renewal periods and this will be influenced by management's policy to standards of appearance and comfort within the establishment. Nothing lasts forever and there is a choice of getting as long a life as possible out of the article or of planning a renewal period.

In the first case, repairing, patching, renovating etc, may become necessary with their attendant costs and inevitably there will be some reduction in appearance and some articles will lose their first comfort as well, eg a mattress may retain maximum comfort for 5–7 years and later still remain sufficiently comfortable to be used.

In the second case, consideration has to be given to the useful life of an article in a given situation; this will determine the renewal period. Consideration should also be given to the use which can be made of the article afterwards; it may be used in staff quarters, in offices or offered for sale. The renewal period will vary with the particular article or surface, the conditions to which it is subjected, the maintenance given and the standards required, but the renewal or replacement date should be projected at the time of purchasing.

When buying, a knowledge of the British Standards Institution and its kite mark and the mark of the Council of Industrial Design and the Public

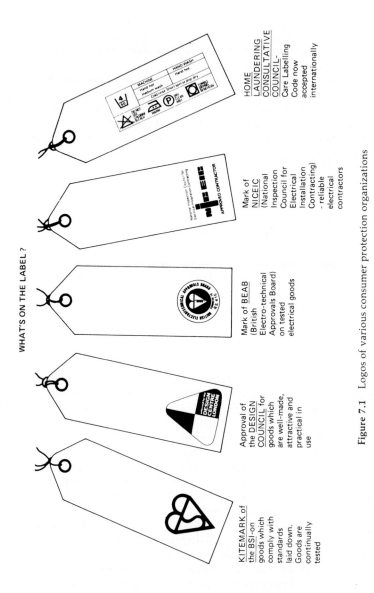

WHAT'S ON THE LABEL ?

KITEMARK of the BSI-on goods which comply with standards laid down. Goods are continually tested

Approval of the DESIGN COUNCIL for goods which are well-made, attractive and practical in use

Mark of BEAB (British Electro-technical Approvals Board) on tested electrical goods

Mark of NICEIC (National Inspection Council for Electrical Installation Contracting) - reliable electrical contractors

HOME LAUNDERING CONSULTATIVE COUNCIL- Care Labelling Code now accepted internationally

Figure 7.1   Logos of various consumer protection organizations

Authority standards (see Fig. 7.1), as well as a knowledge of relevant legal requirements, eg Fire Precautions Act 1971, is advisable.

Information regarding new materials, processes and ideas for the various sections of the industry etc, may be obtained from brochures, trade journals and sales representatives, but it is wise not to accept everything read or heard without verification.

Well selected, good quality articles such as floorings, furniture, furnishings, equipment and supplies may mean less maintenance and enable staff to work more quickly and efficiently, resulting in a saving of man-hours, so good buying may lead to a saving in the wages bill. Standardization may also lead to a saving in money because articles may be interchanged and less stock required. Further saving may be possible by buying one large quantity at one time rather than several smaller amounts, as a better discount may often be obtained; against this, capital is lying idle, storage space is necessary and, in the case of cleaning agents and guests' supplies, they may deteriorate or become out of fashion (see p. 119).

Where appropriate, colours should be seen in natural and artificial light and due consideration given to wastage of material where large patterns are used for fabrics, carpets and wall coverings. Most carpet is available in various widths and careful choice of widths prevents wastage. It may also be economical to buy the same pattern in varying qualities for different areas.

Stock books and monthly consumption sheets are a guide when reordering consumable goods, as quantities used in a given period are then known. Standing orders are sometimes given for certain articles but these can lead to over or under stocking.

Buying may be carried out through:

retail shops, some of which may have contract departments, providing products and services for the contract (non-domestic) interiors;
cash and carry, a disadvantage of which is that provision must be made for transport;
wholesalers;
manufacturers, who may stipulate a minimum quantity to be bought;
a central buying department, from which the individual establishments in a large group may order.

Before an order is placed, evaluation and, in some cases, proving by test are often necessary and the only satisfactory way of ensuring that similar items are being compared is to draw up specifications.

Following the drawing up of the specification, firms should be asked to submit quotations by a certain date. Those outside the price range should be rejected and samples of those within the price range requested. Tests may be carried out on the premises or samples may be sent away to testing houses.

Specifications will differ from article to article; the following table gives an example of a specimen specification for a carpet:

Constructional requirements	
Construction	Wilton
No. of colours	2
Composition of surface yarn	80%/20% Wool/BriNylon
Height of pile (above backing)	10 mm
Backing material	cotton/jute
Performance requirements	

The materials will be subjected to a number of tests including:
    Dynamic loading BS 4052 – loss of pile height after 1000 impacts not more than 30%
    Compression recovery BS 4098 – recovery not less than 80%	

The following table shows the results of some tests on carpet samples:

Sample	No. of rubs	% compression recovery	% loss in pile height	pile weight	Composition		Price
					Wool	Nylon	
1	100 000	78.9	9.8	1722	71.5	28.5	£x
2	44 000	67.8	15.8	811	72.7	27.3	y
3	90 000	73.8	16.1	1187	77.7	22.3	z

All goods should be checked on delivery for quantity, quality and condition, since the sooner complaints are made the better attention they will receive. For security reasons, goods should be sent to the correct storage place as soon as possible.

B

CLEANING

8

CLEANING EQUIPMENT
AND AGENTS

Only 5 to 10% of the cost of cleaning is spent on cleaning equipment and agents (the rest being labour), but they play a major role in the cleaning process. The *housekeeper* should endeavour to provide those which make cleaning easier for her staff, save time and obtain a satisfactory result.

Manual equipment

Brushes

Brooms and *brushes* may be used for removing dust (ie for dry work) from a variety of surfaces – floors, walls, upholstery, clothes etc – and may have bristles of animal, vegetable or man-made origin. Cobwebs may be removed as well as dust from cornices, ceilings and high ledges by the use of a wall broom, the head of which is soft and the long handle made of cane.

In the past a soft broom was used on hard floors and a stiff one on carpets; however, sweeping raises a certain amount of dust and, with the advent of the more hygienic process of vacuum cleaning, the broom is used less often. Brooms and brushes should not be used in hospitals as it is important that dust does not become airborne and spread infection.

Brushes are more frequently used for the removal of dirt (ie for wet work when the bristles are stiffer than those used for the removal of dust). This may be done by hand, using a scrubbing brush, floor cloth, detergent and hot water, or with a long-handled scrubbing brush (deck scrubber) using detergent and hot water; and for the efficient removal of the dirt, the soiled water must be picked up with a mop or vacuum drying machine. Deck scrubbers are useful for corners and round equipment where mechanical scrubbing is not possible.

A *carpet* or *box sweeper* (Fig. 8.1 overleaf) is used for the removal of surface dust and crumbs from carpets. It consists of a revolving brush between two small dustpans and the brush is motivated by the worker. It is not as efficient as a vacuum cleaner but it is a useful and quiet substitute on occasions.

Mops

Dry mops consist of a head of various shapes and sizes, made from soft twisted cotton yarn or synthetic fibres and attached to a long handle; the

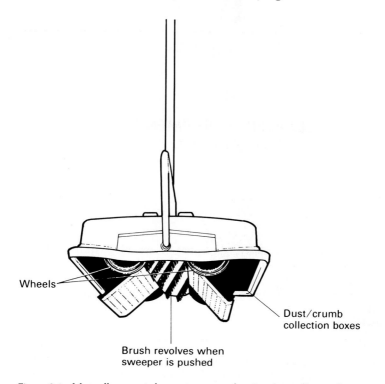

Wheels

Dust/crumb
collection boxes

Brush revolves when
sweeper is pushed

Figure 8.1   Manually operated carpet sweeper showing dust collection boxes open

synthetic fibres are electrostatic and attract the dust. Some cotton mopheads are impregnated with a dressing which causes dust to adhere to the mop more satisfactorily. The large mops are known as mop sweepers (see Fig. 8.2); those with two heads ('v' sweeper) have a scissor-like action and are very suitable for large areas.

When dirty the mopheads should be washed (for safety reasons they should never be wrung by hand), dried and re-impregnated if required and this may be a contract service.

Wet mops or sponge mops are used for cleaning lightly soiled floors in conjunction with a bucket, hot water and detergent. The mop consists of longer, coarser cotton yarn than a dry mop and a sponge mop is another type of wet mop. Both these mops, unless washed well after use, become unhygienic and as with dry mops if wrung by hand there is a danger of accidents.

Polish applicator mops usually consist of an oblong head attached to a long handle; this may be labelled for the type of polish used and the mophead then is not usually washed but replaced as necessary.

Squeegees are used to remove excess water from the floor and smaller ones are used in window cleaning.

Cloths

Dusters and *mitts* are used for the collection of dust from hard surfaces and are usually made of soft cotton or short-life material and mitts may be

Figure 8.2 Mop sweeper

impregnated. Dusting is only an effective method of the removal of dust when the dust is actually collected on the duster. This entails the duster being used in the form of a pad with no loose ends to flick the dust about. Damp dusting (using a swab, mutton cloth or short-life cloth) may prove more effective on some surfaces and is the only method used in hospitals. Dusters should be washed frequently. *Note* The yellow ones have a very loose dye.

Rag may be obtained from the linen room or bought by the sack; it is used for applying polish and when dirty is thrown away.

Wet cloths should be absorbent and of a manageable size so that they can be wrung out by hand. They should be washed and dried after use to prevent them becoming unhygienic. They may be colour coded according to the area or the purpose for which they are to be used.

Swabs may be of mutton cloth or other soft, absorbent material. They are used for wet work above the floor, ie washing paint, baths, wash basins, etc. A short-life cloth, eg J cloth, is equally suitable but not so absorbent.

Floor cloths are made of coarser cotton material than swabs, and are used for WC pedestals and floors when the use of a kneeling mat is advisable.

Chamois leathers were originally skins of chamois goats, but now they are usually skivers, ie split skins of sheep or simulated skins. They are used wet for cleaning windows and mirrors, but they are also used dry as polishing cloths for silver. They should be washed when necessary, and rubbed when dry to soften them. As they are expensive they are only issued as required for special jobs.

Scrim is a loosely woven linen material which, because of its absorbency and not leaving linters, is often used instead of chamois leather for cleaning windows and mirrors.

Dust sheets are made of thin cotton material, about the size of a single sheet, and may be 'discards' from the linen room, eg thin curtains and bedspreads. They must always be kept clean and are used for covering furniture, stored articles and during spring cleaning.

Druggets are made of coarse linen, fine canvas or clear plastic and may be in the form of a 'carpet square' or a runner. They are used to protect the floor during bad weather and during redecoration.

Hearth and *bucket cloths* must be used clean and may be used to protect the carpet or flooring when a fireplace is being cleaned or if there is a likelihood of water being splashed when a bucket is being used.

Containers

Buckets (pails) are normally made of plastic these days because they are lighter in weight, much quieter in use, and very much easier to clean than galvanized iron ones. Mop buckets on castors with wringer attachments are still usually made of galvanized iron (see Fig. 8.3). Twin buckets on a low trolley enable the floor to be rinsed more effectively.

Polish applicator trays are used when applying liquid polish to a floor with a polish applicator mop. They should be clearly marked with the type of polish.

Spray bottles may be used to apply a fine spray of water or cleaning solution as required.

Dustpans are used in conjunction with a brush for the gathering of dust. Formerly they were of metal but now plastic ones are more usual and, in order to be effective, the edge in contact with the floor must be thin and flat.

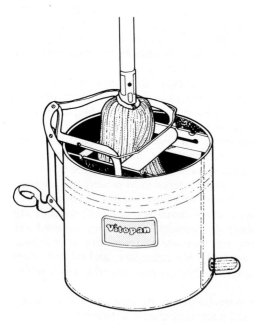

Figure 8.3   Mop and wringer with foot pedal

Dust bins are often kept on the back stairs, in the maids' service rooms or other convenient places. They used to be made of galvanized iron and were very noisy but now they may be of rubber composition or in the form of refuse sacks, which are of strong disposable paper or plastic and attached to a stand. Bins should be emptied frequently and kept clean and sacks should be removed when full.

Sanibins are small metal or plastic containers with lids, found in toilets for the collection of soiled sanitary towels. The bins must be emptied frequently and kept clean, and for hygienic reasons paper bags are often provided for the wrapping of the soiled towels. In some places, incinerators have been installed to replace sanibins and these burn the soiled towels, leaving just a small amount of ash which has to be removed during cleaning. Alternatively, there may be a container with a germicidal fluid into which the soiled towels are placed. The containers are on loan and changed regularly in accordance with the requirements of the establishment.

Housemaids' or chambermaids' boxes were originally made of wood or metal but nowadays are made of plastic. They consist of a box with a handle and a fitted tray, and are used by maids for carrying small items, eg toilet soap, polishes, abrasives, cloths, etc and a plastic bucket is a fair substitute.

Trolleys may be used when a maid's section is all on one level, the corridor wide enough and there is sufficient storage space. A trolley (see Fig. 8.4) is a large fitted conveyance, which as well as holding the items mentioned for the housemaid's box, has a bag for soiled linen and one for rubbish, shelves on which clean linen and other accessories are carried, and also a step on which the vacuum cleaner can rest. In order to prevent pilfering of such articles as soap, tea and coffee bags, sugar, book matches and other 'give-aways', it is not wise to have them displayed on the top shelf.

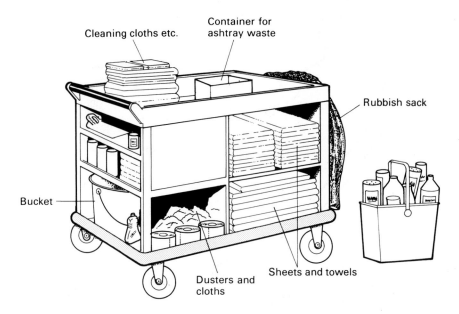

Figure 8.4 Housemaid's trolley and box

Mechanical equipment

Vacuum or suction cleaners

Vacuum cleaners remove dust and other loose particles from hard or soft surfaces by suction and some, in addition, have brushes to aid the collection of dust. The dust is collected into a container which may be enclosed within the body of the machine (*cylindrical* and *cannister models*) or on the outside in the form of a bag (*upright models*). (See Figs. 8.5 and 8.6.)

When the cylindrical and cannister models are in use the hose through which the suction takes place is always attached and the different heads are easily changed, making the vacuum cleaners more adaptable than the upright models which are more suited to large carpeted areas.

Cylindrical vacuum cleaners have no rotating brushes and work by suction only; the term *'suction cleaner'* refers to this type of cleaner. A filter/diffuser may be fitted to the outlet which removes fine dust and micro-organisms from the flow of air from the outlet. In the National Health Service suction cleaners for use in patient and high-risk areas have a triple filtration system (paper bag, cloth bag and filter/diffuser) and the system must conform to BS 5415 and have 60 per cent efficiency. The filter/diffuser also

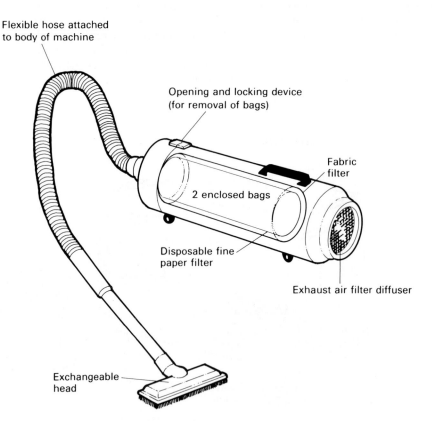

Figure 8.5 Cylindrical vacuum cleaner showing filters

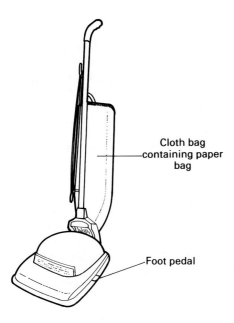

Cloth bag
containing paper
bag

Foot pedal

Figure 8.6 Upright vacuum cleaner

reduces air disturbance and noise. In hospitals the noise should be less than 70 dB.

Small suction cleaners are available for upholstery and carpet edges etc, and these may be carried by hand, eg dustette, or strapped to the back of the operator. There are large cannister-type suction cleaners which may be adapted to collect either dust or water and which are often referred to as *wet/dry pick-ups*. *Suction driers* (vacuum drying machines) only pick up water or dirty solutions from a floor surface.

It is essential that the bag or container into which the dust collects is not allowed to become overfull, or a strain is put on the motor. Damage to the cotton bag should be avoided and it should never be washed or it will allow dust as well as air to pass through it. Fluff and threads should be removed from the brush and wheels; worn brushes and loose belts should be replaced and wheels may need oiling.

It is possible to have a centralized vacuum cleaning system built into an establishment, where ducts carry dust direct to a basement dust room and the spread of dust and micro-organisms is avoided. There are outlets from the ducts into which the cleaning operators insert a hose, to which they attach a suitable nozzle for the particular job. This system is better when built into a new building, as afterwards it is expensive to install; the storage of hoses may be a problem and unless lightweight hose is used it is heavy for women to operate, but there are no frayed flexes and no individual machines to go wrong or to be emptied.

Scrubbing/polishing machines

Scrubbing machines consist of one large or several small brushes which revolve and scrub the floor; the water and detergent are released from a tank attached to the machine. With suitable brushes these machines can be used for shampooing carpets, polishing, spray buffing, spray cleaning or polishing floors.

Combined scrubbing/polishing machines are used in many establishments as the machine can then be put to greater use (see Fig. 8.7). In some circumstances coloured abrasive nylon pads replace the scrubbing brushes. For normal speed machines:

beige pads are used for buffing
green pads are used for scrubbing
black pads are used for stripping

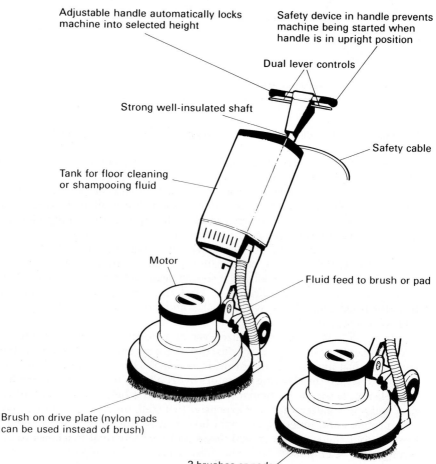

Figure 8.7   Floor maintenance machine (scrubber/polisher/carpet shampooer)

The lighter the colour pad used, the less abrasive action occurs.

These machines may be with or without suction to pick up the soiled water when used for scrubbing, and when without suction a mop or vacuum drying machine (wet pick-up) should follow the use of the scrubbing machine. After use, pads and brushes should be removed and cleaned and never left on the machine.

Hot water extraction machines

These are machines with no rotary action carrying a tank of hot water and detergent, which are used for deep cleaning carpets. The hot water and detergent are 'shot' into the carpet with high-pressure spray nozzles, the dirt is flushed to the surface and it and the soiled water are removed by suction into a container on the machine. Solvent extraction may be used for upholstery and curtains. (See Fig. 8.8.)

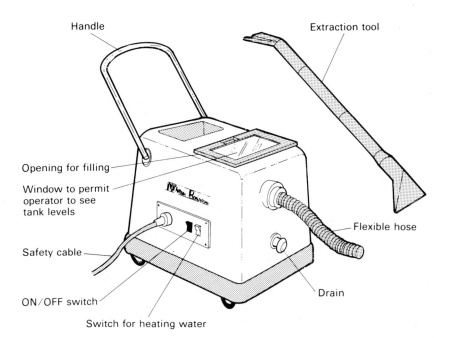

Figure 8.8   Hot water extraction machine for carpets (spirit extraction for upholstery)

Choice of equipment

The *housekeeper* has a great responsibility when choosing equipment since a poor choice can prove less efficient than desired and more costly than it should be. It is necessary to be able to justify its use in terms of saving time and labour, hence the saving of money, and of its efficiency in obtaining a good result.

An expensive piece of equipment not used frequently is a waste of money and so a dual purpose machine might be considered, eg scrubbing/polishing machine. An alternative to buying would be to hire the equipment.

In general when choosing cleaning equipment, the following points may be considered:

safety factors;
ease of operation and cleaning;
work performance (finished result);
saving of time and labour;
durability;
appropriate design, size and weight;
versatility;
manoeuvrability and portability;
noise;
storage;
maintenance and servicing arrangements;
reputation of company;
initial and operating costs.

Care of equipment

Having selected and bought good equipment it is up to the *housekeeper* to see that it is properly looked after. This means that training and good supervision are necessary and she should ensure that the staff

use it properly;
store it correctly;
are given time to clean it;
realize the importance of reporting faults promptly.

Regular servicing is necessary for all electrical equipment; flex and plug defects and unusual working noises should be reported immediately, and unqualified persons should not try their hands at repairs.

When machines break down, repairs may take some time and during this period the lack of equipment may present difficulties. Therefore some establishments have their electrical equipment, eg vacuum cleaners, polishers etc, on hire, the firms being under contract to supply the required number of items in working order.

Equipment may be coded (labelled or distinguished by different colours) for use in different working areas.

Cleaning agents

Dust, being composed of loose particles, is removed comparatively easily by the use of various pieces of equipment; dirt, however, owing to its adherence to surfaces by means of grease or moisture, requires the use of cleaning agents as well as equipment if it is to be removed efficiently, and a knowledge of the different types is important so that deterioration of the surfaces is prevented.

Water

Water is the simplest cleaning agent and some forms of dirt will be dissolved by it, but normally unless it is used in conjunction with some other agent, for example a detergent, water is not an effective cleanser. In fact it does not even wet a surface satisfactorily as its surface tension prevents it from spreading. Hardness in water is another consideration to be borne in mind.

Hardness in water

This is due to dissolved salts of calcium and magnesium (usually bicarbonates and sulphates). Bicarbonates give rise to temporary hardness because they are removed by heating water above 72°C, when 'scale' or 'fur' results:

$$Ca(HCO_3)_2 \rightarrow CaCO_3\downarrow + CO_2 + H_2O$$

Permanent hardness is caused by the sulphates of calcium and magnesium, which cannot be removed by boiling but by water softening processes when temporary hardness is also removed.

Hard water may be *softened* by:

the addition of soda;

certain 'water softeners' based on sodium sesqui-carbonate, eg Boots Laundry Water Softener;

sequestering agents, eg sodium hexametaphosphate (eg Calgon) which form a complex salt with the calcium and magnesium ions, thus removing them from the water but forming no precipitate (scale). These are usually expensive as relatively large quantities have to be used to overcome the hardness;

a water-softening unit, eg Permutit. The hard water passes through a container filled with resin beads containing sodium ions. The resin has a greater affinity for calcium ions (in the hard water) than for sodium ions so an ion exchange takes place; calcium ions are removed from the water and are replaced by sodium ions, which do not cause hardness in water.

Hardness in water will:

have an adverse effect on the efficacy of some cleaning agents, eg soap and soap-based washing powders are wasted and form a scum (lime scale) in the water;

cause premature ageing of fabrics – the fibres degrade more quickly because of the friction produced by the deposits from the hard water, white fabrics tend to lose their whiteness and there can be overmarking of coloureds;

cause scale and fur to be deposited in boilers, pipes and domestic appliances.

Detergents

Detergents are cleaning agents which, when used in conjunction with water, can loosen and remove dirt, and then hold it in suspension so that the dirt is not redeposited on the clean surface.

In order that they may do this detergents require three basic properties:

good *wetting* power to lower the surface tension of the water and enable the surface of the article to be thoroughly wetted;

good *emulsifying* power to break up the grease and enable the soiling to be loosened;

good *suspending* power to prevent redisposition of the soiling.

There are many different detergents available in packets and bottles, each one differing only in the mixture of chemical substances of which detergents may be composed. Detergents may be *soapy* or *soapless (synthetic)*; soap was the original detergent but has in many cases been superseded by the synthetic detergents, although tablets of toilet soap still remain.

The basic ingredients of any detergent are *surface active agents* or surfactants; these are primarily the wetting agents which lower the surface tension of the water and to varying degrees emulsify the grease and suspend the soiling. Each molecule of surfactant has one end which is attracted to water (a hydrophilic or water-loving head) and the other which is repelled by water and attracted by grease (a hydrophobic or water-hating tail). (See Fig. 8.9.)

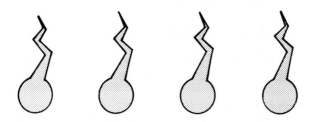

Figure 8.9   The head and tail structure of a typical surfactant molecule. The head is hydrophilic and the tail hydrophobic

When a detergent is added to a drop of water, the surfactant molecules arrange themselves so that their hydrophilic heads are into the water and their hydrophobic tails pushed out breaking the bonds between the water molecules and so reducing the surface tension. The drop of water collapses, spreads and wets the article. Hence the importance of the wetting power of the detergent. (See Fig. 8.10.)

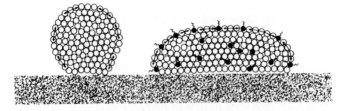

Figure 8.10   When a detergent is added to a drop of water, the spherical structure of the water will collapse because the surface tension has been weakened

Figure 8.11 The removal of grease: the grease is coated with the hydrophobic tails of the detergent molecule and squeezed away from the fibre to satisfy the hydrophilic heads. The droplets of grease are prevented from touching each other by their coating and remain dispersed

As well as lowering surface tension detergents are required to help loosen dirt, particularly greasy dirt. In this case the hydrophobic tails of the surfactant stick into the surface of the grease and the hydrophilic heads into the water. The grease is then lifted away from the article to satisfy the hydrophilic heads, and is emulsified. (See Fig. 8.11.)

The particles of grease and dirt remain suspended in the detergent solution and do not resettle on the clean articles, partly because the hydrophilic heads surrounding the grease and dirt particles are attracted to water and partly because most of the surfactants now carry mild electric charges (ie they ionize) and cause the particles to repel each other.

Surfactants are classified on the basis that when dissolved in water some dissociate into positively and negatively charged particles or ions, while others do not (see table overleaf).

An important consideration in choosing surfactants for a detergent base is that they should be biodegradable, that is, they should break down in rivers and sewage works waters.

To formulate a successful detergent other substances are added to the mixture of surfactants, especially in the case of the powdered 'heavy duty' detergents where increased emulsifying and suspending power as well as other properties are required for the washing of soiled fabrics.

Some of the more frequently added substances are:

alkaline builders, eg soda, borates, silicates and complex phosphates,
– increase pH value of solution and so increase emulsifying powers,
– may constitute 30% of heavily built powdered detergents which may then have a pH of about 9,
– assist in softening water enabling more of the surfactant to remove dirt, sodium tripolyphosphate is a sequestering agent and forms a complex salt with the calcium and magnesium ions of hard water but forms no precipitate (scum);
– when mixed with surfactants the mixture is synergistic, that is, it has

Types of surfactants

Anionic	Cationic	Non-ionic
• carry negative charge on hydrophobic part of the molecule • have a good wetting power • have limited suspending powers • constitute about 20% of most synthetic detergents • include soap and alkyl benzene sulphonate (the most frequently used synthetic surfactant)	• carry positive charge on hydrophobic part of the molecule • counteract anionic surfactants • have limited use in detergents • used in fabric conditioners giving soft handle which is anti-static and less ready to pick up soiling • most important probably are 'quats' (quaternary ammonium compounds) which have germicidal properties	• do not ionize in solution • have excellent wetting powers • have good emulsifying powers • do not lather as well as *anionic* surfactants • are suitable for detergents specially formulated to deal with greasy soiling, eg Drive, and for low-foaming detergents • have high solubility and constitute about 6 to 12% of liquid synthetic detergents but only about 2% of most powdered synthetic detergents • most important are polyoxyethylene ethers and esters

greater efficiency than would be expected when considering the efficiency of the individual substances,

– are not usual in toilet soap and detergents formulated for the washing of woollen and silk articles as the skin and animal fibres are sensitive to alkalis; Sodium silicate prevents corrosion in washing machines and keeps the powder free-flowing;

sodium sulphate is a bulking agent, enabling the detergent powder to flow freely, and may form as much as 20% of the detergent powder;

sodium carboxymethyl cellulose (effective on cotton) and *ethyl hydroxy-ethyl cellulose* (effective on synthetic fibres) are suspending agents and prevent the dirt resetting on the cleaned article;

sodium perborate and *sodium percarbonate* are oxidizing bleaches which decompose during the wash to form hydrogen peroxide. They are most effective at temperatures between 85°–100°C, and will remove tea, coffee and fruit juice stains;

Because of this need for high temperatures a new bleaching agent – *tetra acetyl ethylene diamine* (TAED) – has been developed which works at lower temperatures and is included in New System Persil Automatic. TAED acts as a bleach 'precursor', combining with sodium perborate to produce peroxy-acetic acid (peracid), a different kind of bleach which reaches maximum efficiency between 50°–60°C;

foam or *lather stabilisers*, eg ethanolamides, ensure lather is maintained and have some detergent properties;

fluorescers are whitening/brightening agents, which absorb ultra-violet light and re-emit it as visible light. They maintain whiteness of cellulosics, eg cotton and linen textiles and enhance whiteness best when fabrics are clean;

enzymes are complex proteins which break down organic substances, eg blood, egg, gravy, grass, milk and body soils, such as 'rings' on collars and cuffs, soiling on towels etc. They are biological catalysts and work most effectively about 40°–50°C or during prolonged soaks at lower temperatures. They are inactivated above 60°C;

germicides, perfumes and *dyestuffs* may be added. Perfumes are added to give a clean, fresh fragrance and only in toilet soap do they give a scent in the normal meaning of the word.

Detergents may, therefore, contain many ingredients and this is particularly true of the heavy duty powdered ones, both soap and synthetic. According to the composition of the mixture, each detergent will have advantages for a particular cleaning job.

In order that the washing process should be carried out efficiently, an ideal detergent should:

have good wetting powers so that the solution penetrates between the article and the dirt particles;

have good emulsifying powers so that grease and oil are broken up and to some extent dissolved;

have good suspending powers so that the dirt particles, when removed, are suspended in the solution and are not redeposited on the article;

be readily soluble in water;

be effective in all types of water and produce no scum;

be effective over a wide range of temperatures;

be harmless to the article and the skin;

cleanse reasonably quickly and with minimum agitation;

be easily rinsed away;

be bio-degradable.

Types of detergents

Soap is obtained when fat or oil is treated with an alkali (saponification), it is an anionic surfactant and:

is cheap and effective in soft water,

forms scum in hard water and is difficult to rinse away,

is not effective in acid solutions,

apart from toilet soap has been largely superseded by synthetic detergents.

In hotels 22 g and 56 g tablets of toilet soap are usual and these may be wrapped with the name of the hotel printed on the soap or the wrapping, or unwrapped when careful storage is necessary in order that the soap retains its fresh appearance. The tablets are issued for use in the guests' rooms, private bathrooms, and cloakrooms, and pieces left over from the guests' rooms may be transferred to the cloakrooms for general use, or returned to the manufacturer for remaking, when a slight discount is given.

Synthetic (soapless) detergents (see table overleaf) have replaced the use of soap in many cleaning processes because they are not affected by hard water, have good suspending powers, do not dry with smears and most are stable in acidic or alkaline media (soap is not effective in acid solution). It is for these reasons that as long as the water containing the synthetic detergent remains clean, there is no need to rinse hard surfaces such as walls, floors, etc with clear water.

Liquid synthetic detergents are light duty detergents suitable for washing up and the washing of hard surfaces and lightly soiled fabrics. They do not contain many of the ingredients essential in a heavy duty powdered detergent, such as oxygen bleach or alkaline builders.

Detergents

Toilet soap	*Soap flakes*
• contains no builders • contains perfume, dye-stuffs and possibly antioxidants	• have simplest of all detergent formulae (except for toilet soap) • dissolve easily • used for delicate fabrics washed at lower temperatures
Soap powders	*Liquid synthetic detergents*

• often contain up to 40% builders
• dissolve and lather easily
	• contain approx.
	    20% anionic surfactant,
	    6–12% non-ionic surfactant
	    builders unusual
	    SCMS seldom but may be present
	    water to 100%
	• are near neutral in reaction
	    pH 7 approx.
	• suitable for washing up and the washing of hard surfaces and lightly soiled fabrics

Powdered synthetic detergents	*Biological detergents*

• contain many ingredients of varying types and quantities
• contain approx.
    20% anionic surfactants
    2% non-ionic surfactants
    33% alkaline builders
    9% bleach
    SCMC and brighteners
    20% filler
    15% water
• pH 7–9 approx.
• suitable for heavily soiled surfaces and fabrics
	• are powdered detergents to which enzymes are added
	• used for the removal of organic stains and work most effectively at a temperature of 40°–50°C

Solvent-based detergents	*Sanitizers or disinfectant detergents*
• contain water-miscible solvents, builders and anionic surfactants • pH approx. 12 • used for stripping spirit-based wax floor polishes and the accumulation of grease from kitchen surfaces • used in oven-cleaning formulations	• based on cationic surfactants, mainly the 'quats' (quaternary ammonium compounds) • have good germicidal and antistatic properties • are available as cleaning gels, air fresheners and fabric conditioners • some may be used as residual sanitizers on floors, walls, equipment and working surfaces which come into contact with food

Powdered synthetic detergents contain many ingredients, and by altering these in type and/or quantity a detergent may become more suitable for one purpose or another.

Synthetic detergents will therefore be used for a great variety of purposes, including washing up and the washing of floors, walls, baths, basins and fabrics, and may be bought in bulk or in smaller containers.

Abrasives

Abrasives depend on their rubbing or scratching action to clean dirt from hard surfaces. The extent to which they will rub or scratch a surface depends on the nature of the abrasive material and on the size and shape of the particles.

Glass, sand and emery papers are all forms of abrasives, as are steel wool, nylon web pads, powdered pumice, feldspar, calcite, fine ash, precipitated whiting (filtered chalk) and jewellers' rouge (a pink oxide of iron), the last two being the finest. The use of abrasives will depend on the surface to be cleaned and the type of dirt to be removed; when possible, fine abrasives should be used in preference to coarser ones. On the scale of hardness, which shows talc as 1 and diamond as 10, feldspar has a hardness of 6 and calcite of 3.

Rather than being used alone, abrasives are more frequently used in the form of a finely ground mineral, generally limestone or calcite, as the main ingredient of scouring powders, creams or liquids (see table below).

Scouring powder eg Vim	*Scouring cream or liquid eg Jif*
• contains approx. 80% finely ground limestone to 'scratch' the dirt away chlorine bleach to disinfect surfaces and for stain removal alkaline builders to aid grease removal and dissolving of bleach anionic surfactant to create lather, remove grease and suspend dirt lemon perfume • will scratch surfaces if used too generously • should not be used on paintwork or plastic surfaces • apply with a damp cloth and rinse off	• is milder in action • contains finely ground calcite suspended in a solution of anionic and non-ionic surfactants polyphosphates and soda to maintain alkalinity ammonia, a preservative and a perfume • apply on a damp cloth and rinse off • container needs to be well shaken • can be used on paintwork, acrylic baths etc, when best applied with a sponge and surface wiped gently

Toilet cleansers

Toilet cleansers are crystalline, powdered or liquid and they rely on their acid content to clean and keep the WC pan hygienic. Acids remove metal stains eg limescale. The crystalline cleansers are normally based on sodium acid sulphate, a mild acid which is mixed with an anti-caking agent, often pine oil, which in the past also helped to prevent corrosion of any metal container. The cleansing effect can be improved by the addition of a small amount of acid-resistant anionic surfactant.

Powdered toilet cleansers consist of a soluble acidic powder, chlorinated bleach, finely ground abrasive to help when a brush is used and an effervescing substance which helps to spread the active ingredient throughout the water.

Liquid toilet cleansers may be a dilute solution of hydrochloric acid and should be used with great care because the concentration may cause damage to the surface of the pan, to surrounding areas and to the person using it if the liquid is spilt.

All these toilet cleansers are designed for the cleaning and disinfecting of lavatories and urinals only and should *never* in any circumstances be mixed with other cleansers, because harmful gases are likely to be produced.

Liquid chlorine bleaches, which are alkaline-stabilized solutions of sodium hypochlorite, may also be used to clean and disinfect lavatories and should never be mixed with other lavatory cleansers. The solution may contain detergents which assist the cleansing action and increase the viscosity, so enabling it to adhere to the sloping surfaces of the WC pan (eg Domestos).

Window cleansers

Window cleansers consist of a water-miscible solvent, often isopropyl alcohol, to which a small quantity of surfactant and possibly an alkali, are added to improve the polishing effect of the cleanser. Some also contain a fine abrasive. The cleanser is applied with a cleaning rag and rubbed off with a clean, soft cloth.

Water, or water to which some methylated spirit or vinegar has been added, does the job quite well and much more cheaply but entails more rubbing.

Soda and ammonia

These are alkalis, and are used as grease emulsifiers and stain removal agents. The addition of alkaline salts to surfactants in the formulation of detergents has already been mentioned.

Strong alkaline cleaning agents based on caustic soda in flake or liquid form are available for the clearing of blocked drains, cleaning ovens and other large industrial equipment. Extreme care has to be taken in their use as they are very strong materials with high pH values (1% solution may have a pH of 13.1).

Acids

Acids dissolve metals and are used for the removal of metal stains, such as water stains in baths, hard water deposits round taps and in WC pans, tarnish on silver, copper and brass articles.

Vinegar and lemon (cut or juice) are used for the removal of tarnish from copper and brass and of mild water stains on baths, etc. The acid produces further staining on the metals if it is not washed off quickly and on sanitary fitments it may damage the glaze. More resistant water stains may be removed with stronger acids such as *oxalic acid* or *spirits of salt* (concentrated hydrochloric acid). These should only be used under strict supervision, and in all cases of cleaning the acids must be thoroughly rinsed away or they may harm the surface.

(There is a variety of proprietary substances sold under trade names which are helpful in the removal of hard water deposits.)

Mention has already been made of the acidic ingredients of toilet cleansers and further use of acids in metal cleaning is included on pages 114 and 155.

Paraffin oil is also efficient for the cleaning of baths but owing to its smell is seldom used.

Organic solvents

Organic solvents, usually methylated spirit, white spirit (turpentine substitute) and carbon tetrachloride are grease solvents, and used for the removal of grease and wax from different surfaces. The two former are highly inflammable while carbon tetrachloride is harmful if inhaled, and should therefore never be used in a confined space. (See pages 138–40.)

Aerosol dry cleansers, suitable for use on wallpaper and furnishings, are available.

Bleaches and disinfectants

Bleaches used for cleaning purposes are generally alkaline stabilized solutions of sodium hypochlorite and are useful for stained sinks, WC pans etc, but they should never be mixed with other types of toilet cleansers. They whiten and have germicidal properties and great care should be taken to prevent spotting of other surfaces. A blend of surfactants is sometimes added to increase cleaning power, eg Domestos. Strong solutions of chlorine bleaches corrode or discolour copper, aluminium, silver and stainless steel. Other bleaches are mentioned in connection with the removal of stains from fabrics on page 141.

Disinfectants, antiseptics and deodorants are not strictly cleaning agents but are often used during cleaning operations. The use of disinfectants and antiseptics should be controlled carefully as many have strong smells and their use often suggests illness or bad drains.

Disinfectants kill bacteria; antiseptics prevent bacterial growth and are frequently diluted disinfectants; deodorants mask unpleasant smells either by combining chemically with the particles forming the smell, or by their smell being predominant and may be obtained as aerosol sprays.

Quaternary ammonium compounds (cationic surfactants) are useful bacteriocides and deodorants, but they cannot be used with anionic soaps or soapless detergents.

Polishes

Polishes do not necessarily clean but produce a shine by providing a smooth surface from which light is reflected evenly. They do this by smoothing out any unevenness on the surface of the article, as in polishing metals, or by providing a very smooth protective layer as in floor and furniture polishing. Their composition will therefore depend on the surface for which they are intended and general principles in their use are:

apply to a clean surface;

use only when a satisfactory finish cannot be obtained without it;

use the smallest possible amount to obtain the desired finish;

remove all traces of the polish;

ensure the correct type of polish and polishing methods are used for the particular surface to be polished.

Metal polishes

These remove the tarnish resulting from the attack on the metal by certain compounds in the air and some foodstuffs. They consist of a fine abrasive (generally precipitated whiting and jeweller's rouge) which, when rubbed on the surface of the metal, provides friction to remove the tarnish and produce a shine.

In many proprietary metal polishes, eg Brasso, Silvo, the abrasive powder is mixed with a grease solvent, and in some cases an acid, to help in the removal of the tarnish. The polishes may be liquid, foam, 'long term' or impregnated wadding and are normally formulated for hard, eg brass, or soft, eg silver, metals.

(For further details of metal cleaning see pages 154–6.)

Furniture and floor polishes

Furniture and floor polishes are protective finishes which provide the surface of the furniture or floor with a thin layer of wax or resin. This layer gives protection against abrasion, absorption of spillages etc, and a smooth surface from which light may be reflected to give a shine or sheen.

In general the requirements of a good polish, whether for furniture or floor, are that it should:

give a hard dry finish to ensure maximum protection;

give an easily cleaned surface;

not mark easily;

reduce costs of cleaning and maintenance;

not smell unpleasant.

Furniture polishes (see table opposite) contain:

a special blend of waxes, eg beeswax, carnauba, paraffin wax, or ozokerite
spirit solvent, to soften the wax,
silicones, in some cases, which make the wax easier to spread, give a harder and more lasting finish, improve resistance to heat, moisture and sunlight.

The polishes may be paste, cream or liquid, depending on the wax content and some may be 'spray-on'. It is important that the right polish is used for the surface to be treated and a great deal of modern furniture has a synthetic resin finish which does not require polishing.

Floor polishes

Floor polishes are spirit or water based. *Spirit-based floor polishes* (see table opposite) contain a blend of mainly natural waxes dispersed in a spirit

Types of furniture polishes

Paste furniture polishes	Cream wax polishes
• contain 25–30% wax • may include silicones • are applied sparingly with a rag and rubbed up well with a soft cloth • may give a sticky finish if too much applied at one time and if not sufficiently well rubbed up • leave a layer of wax as spirit evaporates • build-up of layers gives high gloss in time • are suitable for antiques and other pieces of wooden furniture where the shine is dependent on the layers of wax • also suitable for wood panelling and, if not too high silicone content, for wood and cork floorings	• are emulsions of a blend of light coloured waxes in water and oily solvents • contain approx. 20% waxes • have some cleaning action because the emulsifying agent is usually a detergent • may or may not contain silicones • are best applied with a damp cloth and rubbed up immediately • build-up of wax layers slower than with paste polishes • are suitable for most types of wooden furniture • are liable to show in ornate parts of furniture and so these require extra care
Liquid furniture polishes	Spray-on furniture polishes
• contain 8–12% wax • high percentage of solvent gives some cleaning action • are good for removal of food stains, drink rings and finger marks • should be applied with soft cloth and the resultant haze wiped after a few minutes • are most suitable for furniture which already has a shine eg French polished furniture	• contain approx. 8% wax so are similar to liquid wax polishes but alteration in the solvents necessitates the surface being wiped immediately – not left to dry • are expensive and may be wasteful if not used carefully • may be sprayed directly onto the surface or on to a cloth when dusting, particularly for small areas • cleaning and polishing may be combined in one operation • if used on hard surfaces, eg glass, the build-up of wax may make the surface smeary

solvent; they may be paste or liquid and are suitable for use on wood, cork and magnesite floorings, ie those harmed by water. With the introduction of the newer floor materials, spirit-based polishes now account for only about 15 per cent of all floor polishes made.

Spirit- (or solvent-) based floor polish		
Waxes eg natural – carnauba – ozoberite synthetic – polyethylene	Solvent eg white spirit Freon 11 or 12	Additives eg perfumes dyes silicones

Water-based polishes (see table overleaf) are emulsions in which fine particles of natural and synthetic waxes are dispersed in water and are suitable for use on thermoplastic, rubber, PVC and asphalt composition floorings, ie those harmed by spirit. They may also be used on sealed wood, cork and magnesite. Water-based floor polishes account for about 80–85 per cent of all floor polishes made and are always liquid but may be fully buffable, semi-buffable or dry bright.

Water-based (emulsion) floor polishes				
Waxes eg carnauba montan polyethylene	*Water* and emulsifying agent	*Polymers* eg polystyrene acrylate styrene acrylate co-polymer vinyl	*Plasticizers*	*either* alkali-soluble resins *or* metal-complexed polymer

The types and quantities of the waxes and polymers in either type of polish are chosen to provide a protective finish which will:

withstand heavy wear;
withstand spillages and washings;
provide slip retardance;
give a good gloss;
give an easily cleaned surface.

In *spirit-based polishes* the waxes are mainly natural ones but polyethylene wax is used in small amounts because of its durability, gloss and slip resistance.

Silicones, if present, are only in small quantities as they tend to make the floor slippery but they do help to form a hard glossy finish with water repellancy.

The consistency of the polish depends on the amount of wax present.

Paste polishes contain 25–30% wax
Liquid polishes contain 8–12% wax

The latter is easier to spread and there is less 'drag' on the polishing machine. In each case buffing is necessary to produce the shine and avoid a slippery finish. The polish is removed when necessary by loosening the build-up of wax with spirit or a solvent-based detergent, a nylon web pad under the machine will help loosen the wax; the loosened wax should be picked up as soon as possible with a damp cloth, mop or wet vacuum pick-up; delay allows the spirit to evaporate and dirt to be redeposited.

Water-based polishes are always liquid and the amount of wax determines the amount of buffing required to give the shine.

Fully buffable polishes are wax rich, 45–60% wax and 20–40% polymers
Semi-buffable polishes contain 25–40% wax and 45–60% polymers
Dry bright polishes are polymer rich, 5–15% wax and 50–70% polymers

The alkali-soluble resins provide weak points in the water-based polishes, enabling them to be marked by detergent solutions and liquid stains and removed with alkaline detergents.

More frequently now a metal-complexed polymer (often zinc-crossed linked vinyl polymer) is included. This shields the break points in the polymer chain against the penetration of detergent solutions. The result is a tough, durable and water- and stain-resistant finish.

Wetting agents are added to make the polish easier to spread and plasticizers prevent any powdering or flaking off due to the brittleness of the polymer.

Water-based polishes are removed when necessary by loosening the build-up of synthetic finish with a stripping agent (a highly alkaline detergent, pH 10–11); a nylon web pad under the machine will help loosen the build-up of polish. The loosened polish should be picked up with a mop or wet vacuum pick-up and the floor rinsed; vinegar should be added to the final rinse to neutralize the stripping agent.

Diluted emulsion polish or water-based ready-to-use compounds may be used for spray buffing a floor (see page 164).

Floor seals

Floor seals are semi permanent finishes applied to a flooring to render it impermeable, to protect its surface from dirt, stains and other liquids and to provide an easily maintained surface. The floor surface must be clean and dry before a seal is applied or it will not 'key' to the surface. There are five main types of seals:

1 Oleo-resinous seals are penetrating and coat the surface, they are used on wood, cork, magnesite and wood composition floorings.
2 One-pot (ready to use) plastic seals do not penetrate the surface and are finishing seals. They consist of either urea-formaldehyde resin with an acid catalyst or of polyurethane, and are used on the similar surfaces to (1).
3 Two-pot plastic seals contain either urea-formaldehyde or polyurethane as the base with an accelerator or hardener. They are used on similar surfaces to (1).
4 Pigmented seals are either synthetic rubber or polyurethane coloured seals and are used on similar surfaces to (1).
5 Water-based seals provide a plastic 'skin' over the surface and contain acrylic polymer resins and plasticizers to assist flexibility. They may be used on PVC, thermoplastic, rubber, porous linoleum, terrazzo, marble and bitumastic floorings.

Sealed floorings may be polished with water-based floor polishes.

Choice of cleaning agents (consumable items)

With the variety of cleaning agents on the market the *housekeeper* should remember that a great deal of time, effort and money can be wasted by wrong choice as well as possible deterioration of articles and surfaces. Cleaning agents are chemicals and the *housekeeper* should have some knowledge of cleaning science if they are to be chosen and used correctly. The following points may be taken into consideration when choosing cleaning agents:

type of soiling
composition
ease of use
saving of time and labour
possible damage to surface
toxic or irritating to the skin
smell
versatility
packaging
storage and deterioration
cost.

Storage and replenishment of cleaning agents

Stores may be obtained by staff:

- going to a main store run by a storekeeper; the cleaning materials are issued to the individual maid or cleaner at set times when the rule of 'new for old' or 'full for empty' may be applied;
- making out requisition lists which are handed in for the *housekeeper* to countersign and the items are collected later from the stores by a porter or maids;
- going at set times of the day for their replenishments or renewals to a housekeeping store kept under lock and key, which is the responsibility of an assistant housekeeper;
- who may have their stock of supplies 'topped up' by a houseporter several times a week;
- who may collect their box of cleaning materials from the housekeeping stores or even the linen room daily en route to their sections; when the maids finish their work the supplies are returned to the same place to be replenished for the next day.

From the great variety of cleaning agents available the *housekeeper* will normally supply a maid with a suitable:

detergent
scouring liquid
WC cleanser
mirror cleanser
furniture polish
air freshener

In the housekeeping stores there will be other cleaning agents, eg lime stain remover, available when required and a variety of cloths. The usual cloths needed by a maid are:

swab, sponge or disposable cloth ⎫
floor cloth   ⎬ for wet work

basin and glass cloths ⎫
dusters   ⎬ for dry work
polishing rags   ⎭

When buying cleaning agents, powdered items and liquid detergents etc may be bought in bulk; this involves the issuing of small quantities in suitable containers, when it is possible with careless handling for wastage and mess to occur.

Although there may be an economy of money when buying in bulk, wastage of materials can occur, and there is much more time involved in the issuing of broken quantities. All containers should be clearly labelled. New types of cleaning agents should always be well tried out in small quantities before a bulk order is placed.

Toilet paper is ordered by the gross and often arrangements are made for deliveries to come automatically, unless otherwise requested. When ordering, the type of fitment must be remembered and these may be for interleaved or roll-type paper. The paper may be thin and smooth or soft tissue, and in many instances both kinds are provided in the same toilet.

In all cases involving storage, rotation of stock should be practised, and items which are little used should obviously be bought in smaller quantities.

Where the items are requisitioned from a main store, a stock list kept by the *housekeeper* is not so important; but where deliveries are made direct to the housekeeping department, a much more careful check of stock is necessary in order to prevent waste and running out of stock. The frequency with which stocktaking is done varies from establishment to establishment.

Where items are bought in bulk, unless there are large scales, actual stock cannot be taken, so in a housekeeping store the stock of these items is an estimated amount.

Part of a stores sheet could be as shown in the table below.

Part of a stores sheet

Item	Unit	Stock in hand	Receipts	Total	Less issues	Book stock	Actual stock	Discrepancies
Air freshener	tins							
Detergents liquid	litres							
Dusters	each							
Mirror cleansers	bottles							
Polish furniture	tins							

Comparison of the book and actual stock may lead to the discovery of discrepancies which should be investigated. These may be due to poor bookkeeping, careless issuing or pilfering.

In addition to the cleaning cloths and agents there will be other items required in the department for the use of the *guests* or staff, especially in hotels. The following is a list of some of these items which may be kept in the housekeeping stores:

toilet soap

toilet paper

drawer lining paper

writing paper

coat hangers

disclaimer notices

laundry and dry cleaning lists

ash trays

spare electric light bulbs

candles

electric blankets

electric razors ⎫

electric tooth brushes ⎭ for hire

book matches

paper tissues

impregnated paper shoe shiners

brochures

'do not disturb' cards

tooth glasses

hot water bottles

bedpan and urine bottle

other accessories put into guests'
rooms according to house custom
('give-aways' or guests' supplies)
eg toiletries, sachets of detergent,
shower caps, mending kits,
bottled water

A small supply of toothbrushes (including those for electric holders), toothpaste and sanitary supplies may be kept for sale to guests when required. These may be kept under lock and key by the *housekeeper* along with supplies for 'do-it-yourself' tea and coffee making.

In hotels maids have a *service room* where they keep their equipment, cleaning agents and other necessities for their work, and it should be large enough to 'house' the trolley if used. In other establishments articles may be stored in a cupboard.

In any establishment some or all of the following may be provided:

a sink with hot and cold water and a draining board;

a floor sink, similar to a shower tray, with the taps so positioned that buckets and other containers (eg floor mopping and scrubbing equipment) can be filled and emptied easily;

electric or gas water boiler, or large kettle with some means of heating it;

table and chair;

cupboard or shelves for early morning tea-trays and china;

cupboard for floor linen stock;

rail for drying tea towels and dusters;

storage space for cleaning equipment and cleaning agents;

rubbish bin or disposable paper sack;

space for maid's trolley if used.

9

THE LINEN ROOM

The linen room is the central depot for all linen and from it sufficient clean articles, in good condition, are distributed throughout the house.

Linen, in this context, means launderable articles, but the linen room staff may also handle blankets, curtains and loose covers as well as articles for dry cleaning. (*Note* 'Linen' is the only fibre name which also applies to a fabric. Cotton, nylon, terylene etc, are fibre names and the fabrics made from them may be, for example, cotton sheeting, cotton damask, nylon damask, nylon brocade, terylene net etc, or in general terms cotton fabrics, nylon fabrics, etc.)

When one considers that even for a small establishment many hundreds of articles are necessary for the bedrooms alone, it will be realized that the linen keeper (under the supervision of the *housekeeper*) has a great responsibility for the control of this stock.

There will be establishments where a great deal of linen is handled, eg in hospitals, or in hotels where there is much banqueting and daily re-sheeting. In these places there may be a linen keeper with several assistants. In small establishments, where less linen is handled, the work may be done by the general assistant or a linen maid who may even be part-time.

In hostels for various types of people and similar places, where complete re-sheeting for the residents is usually once a fortnight (ie one sheet, one pillowslip per week and residents may provide their own towels), very much less linen is handled and so a linen maid may do the work part-time.

Whenever linen hire is used there is a saving of work in the linen room (see page 133) so fewer staff are required.

The linen keeper or person in charge of the linen room, except in hospitals, is responsible according to house custom for the issue of all linen, the sorting and despatch of the soiled linen to the laundry, the checking on its return and for its general standard. According to house custom she keeps as strict a control as possible over the exchange of soiled for clean linen. She should be capable of being firm with the laundry manager over such difficulties as careless laundering and losses, and she should keep the record books accurately and efficiently.

Linen-room work

Inevitably there are variations in the work of the linen room in the different types of establishments but there are many points to be considered generally.

The hours the linen room is open will vary; 8 am until 5 pm for a hospital or large hotel is usual but in other cases there may be a set time, possibly twice a day, always remembering that the linen room may be required to be open seven days a week.

Security

When the linen room is closed the door should always be kept locked and the key taken to a responsible person according to house custom. No unauthorized person should be allowed access to the linen room but an authorized person should have access to some linen in the case of emergency outside the normal working hours. In an hotel the duty manager or night porter may have a small store or he may remove items from the linen room and leave a note with details of what has been removed. The rule of 'clean for dirty' is considered the best way of keeping control on linen with regard to losses and careless use but is not the only way of exchange.

Exchange of linen

This may take place by soiled linen being:

- directly exchanged for clean over the counter by maid, cleaner, waiter or house or kitchen porters;
- listed and bundled, then taken to the linen room by the maid, house or linen porter at a set time each day and the clean linen collected later in the day;
- dispatched down a linen chute and the floor stock of clean linen made up later in the day by the house or linen porter ie 'topping up' of floor stock;
- collected frequently from the corridors or maids' service rooms by the linen porter and the floor stock of clean linen made up later in the day by the house or linen porter.

In hospitals soiled linen is sent straight to the laundry from the wards etc, and returned via the central linen room.

In a large hotel it is not practical for a maid to make several journeys to the linen room to exchange her soiled linen directly over the counter, so she has a supply of linen in reserve, usually enough to re-sheet her section. This reserve is kept in a floor linen cupboard under lock and key and is made up each day after the soiled linen has been sent to the linen room. The usual checks are not possible with a chute but much time is saved and there is less likelihood of pilfering and misuse of the soiled linen.

Spot checks of all areas likely to harbour soiled or damaged linen, eg bars, staff changing rooms, should be made frequently.

Dispatch

Soiled linen should be sent to the linen room as soon as possible for dispatch to the laundry because to leave it lying about invites misuse and, if in a damp condition, iron mould and mildew can occur, and both these stains need special treatment for their removal. Badly stained articles should be sent to the laundry separately from other soiled linen, so that they may receive special attention, which is an added expense.

As far as possible, similar items are placed in one basket, and care must be taken that no tapes or corners are left hanging out as they may get torn or badly marked. Linen is usually transported in wicker baskets (size approximately 75 cm × 45 cm) firmly fastened by straps, but sometimes canvas bags or vinyl hampers are used.

A list of soiled articles sent to the laundry is given to the van driver when he or she collects the baskets. A duplicate is kept in the linen room. At the same time as soiled linen is picked up, baskets of clean linen are delivered and care must be taken to keep baskets of soiled and clean linen separate as mistakes can easily happen. The frequency with which a laundry collects and delivers depends largely on the amount of linen being sent. For large hotels in a town, pick-up and delivery happen every day (except for weekends and bank holidays) and the time between collection and return of the articles may be only 48 hours.

Inspection

Clean linen is removed from the baskets as soon as possible after delivery has been made. The articles are counted on to the inspection table. 'Shorts' are noted and entered on the next day's laundry list. In some large hotels the laundry sends a checker to count, with one of the linen room maids, the soiled linen going to and the clean linen returning from the laundry. In this way time is saved and there is less likelihood of misunderstandings arising over numbers of articles sent to and returned from the laundry. Ideally, after the clean linen has been counted, and before it is put on the shelves, it should be inspected for:

repairs,
stains,
very bad creasing,
articles belonging to other hotels.

This means that each article has to be opened out and, if necessary, put aside for mending or for return to the laundry for exchange or re-wash. Badly torn articles or 'light' linen, ie linen worn thin, are put on one side for the housekeeper or linen keeper to condemn or discard, to enter in the 'condemned' book and later to mark off in the stock book.

Skilled workers can inspect large articles alone by holding them up to the light, or by placing them flat on the table, but it is often quicker if two linen maids work together, and when inspecting large numbers of small articles, eg napkins, it is less tiring if the maids are provided with chairs.

If inspection is carried out thoroughly on all articles, assuming there is sufficient staff, a high standard of linen is maintained, and the chance of a guest having a napkin with a stain on it or a sheet with a hole in it is less likely. It means also that the linen has a longer life owing to a 'stitch in time saving nine'.

Work study experts have observed that in some situations the 'inspection' of the linen could be omitted. They suggest that it could be made the responsibility of the users, eg waiters and room maids; however, housekeepers and linen keepers, who take pride in their linen, argue that if inspection is omitted then standards must be lowered because careless workers will either use the damaged article, or put it with the soiled linen so that it goes to the laundry again and room maids will waste much time when there is no floor linen stock, in going to exchange faulty articles. Time and labour may be saved in the linen room by the laundry returning the linen in packs, a single pack consisting of two sheets, two pillowslips and towels according to house custom. In this way counting packs is easier than counting individual articles and inspection time is cut in the linen room.

Where a laundry operates on the premises, the inspection could be the responsibility of the laundry workers and linen for repair could then be kept separate and sent direct to the repair department.

However, each situation must be considered on its own merits and linen standards must be balanced against savings in time and wages. There are establishments which hire their linen instead of buying it and in these cases there is less stock, no inspection and no repairs, and so less work in the linen room. As a result of this fewer staff are needed.

Storage of linen

Storage is important so that the linen may air and rest before re-use (see Fig. 9.1). The shelves on which the linen is stored should be firmly fixed, as the weight on them may be considerable, particularly the weight of large linen table cloths and sheets, and the shelves should be clearly marked for

Figure 9.1   Storage in the linen room

each type of article. They should reach to the ceiling and there should be room to mop or vacuum clean under the bottom shelf. In order for the linen to be kept aired, the room should be warm and the shelves slatted to allow free circulation of air.

During storage linen must be kept free from dust but it is inevitable that where linen is being handled dust and fluff will occur and so all linen should be covered. Linen in constant use may be covered by curtains which draw across the shelves, or stored in cupboards with sliding doors. In the case of less frequently used articles, eg special banqueting cloths, curtains and extra blankets, linen covers (often condemned sheets) may be used to wrap round them.

Linen wears better and lasts longer if it is allowed to rest and so a good stock of linen should be kept and always used in rotation. This means that all freshly laundered articles should be put at the bottom or back of the pile. To make counting easier, linen is stacked with the folds outwards and small articles, eg napkins, are placed in tens and secured with a rubber band, often nine with the tenth wrapped round.

As well as the storage of linen kept in normal use a reserve stock needs to be kept and this is frequently stored in its original packing paper in a cupboard under lock and key. The cupboard may or may not be in the actual linen room and a list must be kept of this stock.

Many articles may be issued new and unlaundered but tea towels and glass cloths are always laundered before issue and a good linen keeper will have some of these stored ready laundered.

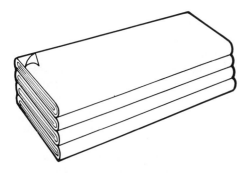

Figure 9.2 Well-stacked pile of sheets

Stocktaking

This is done at specified intervals in order to check the amount of linen, to know when to order new linen and, if possible, to check on losses. It may be done at three or six monthly periods or at any frequency in between. In order to prevent discrepancies it is better if all stock is taken on the same day.

Every piece of linen throughout the establishment should be counted, and the number at the laundry, according to the laundry book, added. Stock is taken by responsible people in each department on the same day, and the lists handed in to the linen keeper who makes up the stock book.

Month _____ Year _____							
Article	Stock in hand	New stock added	Total	Less con- demned	Total	Actual stock at stock- taking	Discrep- ancies
Sheets: single							
double							
cot							
Pillow slips							

Figure 9.3   Part of a sample page from a linen stock book

Each page in the stock book will show the alterations in the linen stock for that particular stocktaking period, and if there are serious losses, the matter should be investigated and control tightened. (See Fig. 9.3.)

Repairs

For economy, mending should be carried out before laundering, but dealing with soiled and perhaps wet articles is not pleasant so mending is done on clean linen. Due to the high cost of labour little hand sewing is done in the linen room but a great deal of machining takes place and thus a sewing machine gets much use and needs to be kept in perfect order. Machines should be dusted and oiled by the operators, and an arrangement should be made for the regular servicing of them on contract. Good light for machining is essential as well as the necessary tools and equipment, such as needles, scissors, cotton and a suitable chair.

Any article not quite up to standard for guests' use in an hotel may be marked for staff or renovated and the last use of all linen is for rag, which will be used for many cleaning purposes. Renovating is not always as economical as used to be thought, owing to the high cost of labour, and so in some linen rooms far less is done nowadays.

Thin places, small holes and cuts frequently occur in towels, table and bed linen, and these are repaired by machine darning. For this work it is necessary for the operator to use both hands, so an electrically powered or treadle machine is essential.

Machine marking may be carried out in the linen room or labels may be sewn or ironed on to articles to denote the name of the establishment or department. Other means of marking include:

marker pens;
iron-on or sew-on labels;
heat-seal machines;

embroidery;
woven.

When marking linen it is usual to mark on the right side of the article and marking may be done on any linen except perhaps waiters' jackets, aprons, kitchen cloths, dusters etc, which are more usually stamped.

Guests' personal laundry

In an hotel it is usual to put a laundry list and sometimes a container, such as a large paper bag, as well as a dry cleaning list in all bedrooms for the guests' personal laundry. The guest is asked to complete the list and to fill in the service required, eg normal or 'express', and the room maid or valet takes the parcel to the linen room. The linen keeper enters the particulars into a guest laundry or dry cleaning book and the van driver collects the parcels.

On its return, the parcel is sent to the guest's room via the valet, room maid or hall porter, according to the custom of the house, and the amount to be charged on the guest's bill is given to the bill office or reception. In some cases dry cleaning is not sent via the linen room but via the valet or hall porter.

Staff uniforms

Articles such as waiters' jackets, aprons and cleaners' overalls, are treated as normal linen room stock, and exchanged over the counter. But where the staff is provided with individual uniforms, this is treated as personal laundry, and may be sent as individual bundles to the laundry and returned a week later. There are certain members of staff (amongst whom may be the assistant

Figure 9.4 Section of the linen room for uniforms, many of which will be sent for dry cleaning

housekeepers) who have their 'dress' dry cleaned periodically at the hotel's expense and in a large hotel there may be a section of the linen room given over to the care of uniforms (see Fig. 9.4).

Linen room work in establishments other than hotels

While many of the points already mentioned concerning the work in the linen room may apply to establishments other than hotels, there are some important differences in hospitals and hostels.

Hospitals

In hospitals, staff who are resident change their own bed linen weekly on clean sheet day and the soiled linen goes from the residences direct to the laundry. The clean linen is then delivered to the residences. Arrangements are made for the exchange of staff uniforms at the central linen room. With regard to the possibility of spread of infection, the problem of nurses' aprons has been overcome by the use of disposable ones.

Soiled linen from the wards, departments etc, is sent direct to the laundry and the clean linen is returned via the central linen room where any repairs may be carried out. In addition to the 'normally soiled' linen there will be fouled and/or infected linen to go to the laundry. This should be placed into waterproof bags immediately it is removed from the bed and the securely closed bags enclosed in a washable outer canvas bag for transit to the central disinfection area. Eventually it will be returned from the laundry in the normal way. (See page 136.) Where a laundry has no central disinfection area the fouled and/or infected linen is sluiced, disinfected and dried at the hospital before being sent to the laundry. Special arrangements are required for linen used by patients with notifiable diseases, eg smallpox, anthrax, typhoid, infective hepatitis etc.

Only in special wards and long-stay hospitals is patients' personal laundry dealt with in the linen room.

From the above it will be realized that in a hospital the rule of 'clean for dirty' is not normally applied, the replacement of clean linen for soiled is more often a 'topping up' system and inspection is kept to a minimum. Some hospitals are fortunate in that they require fewer paid sewing maids as help is given by voluntary workers.

Hostels

In most hostels, 'homes' and university halls of residence, collection and delivery of laundry will be weekly, owing to the less frequent changing of linen. Consequently, unless there is much sewing dealt with in the linen room, it need not be open every day. In many instances the rule of 'clean for dirty' will not apply as the cleaners may be issued with clean linen before stripping the beds. Where the residents change their own beds they are some-times expected to strip their beds and leave the soiled linen for the cleaners to pick up before the clean linen is left. Some halls of residence require the residents to provide their own linen but the articles go to the laundry via the linen room.

Efficiency in the linen room

Launderable linen is required throughout the establishment, and thus the linen room is an essential and important place. Therefore much thought should be given to its situation and planning in order that the work of issue, collection, storage and upkeep of the articles can go on as smoothly as possible.

Ideally, the *linen room* should be situated with direct and easy access for the loading and unloading of linen baskets to and from the laundry, and for the distribution of linen throughout the establishment.

In order that the work should be carried out efficiently the linen room should:

- be large enough for the necessary work to be carried on without over-crowding
- have an easily cleaned floor which will withstand the dragging of baskets and which will not be too noisy
- have walls of a light-coloured washable paint
- have windows, if possible; in any case, lighting should be good and free from glare since many of the articles to be dealt with are white
- have adequate ventilation and heating to keep the stored linen well aired and to prevent mildew occurring

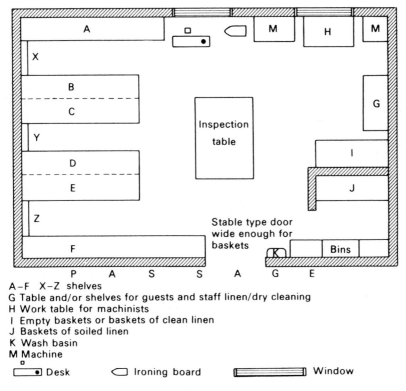

A–F X–Z shelves
G Table and/or shelves for guests and staff linen/dry cleaning
H Work table for machinists
I Empty baskets or baskets of clean linen
J Baskets of soiled linen
K Wash basin
M Machine

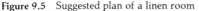
▭•̶ Desk   ◁ Ironing board   ▭▭▭ Window

Figure 9.5 Suggested plan of a linen room

- have slatted wooden shelves to allow free circulation of air
- have a counter or stable-type door over which articles may be exchanged, and to prevent the entry of unauthorized people
- have a door with a strong lock for security reasons, wide enough to take laundry baskets and trolleys
- have a wash basin, soap and towel

In order to carry out this work, the staff will require:

- baskets or bags in which to pack soiled linen
- a table as a working surface, of a colour to contrast with the white linen
- a trolley or floor basket on wheels to save labour
- steps to reach high shelves
- sewing machines for repairing and marking the linen
- an electric iron and ironing board or table
- a suitable table or desk, with drawers for the keeping of record books
- a telephone
- chairs for those who may work seated
- a brush and dustpan or mop sweeper or suction cleaner.

From Fig. 9.5 it will be seen that the principles of work study have been followed as far as possible.

Soiled linen on arrival at the linen room is sorted into bins and to prevent any confusion between soiled and clean linen baskets, the laundry driver delivers the clean ones to I and picks up the soiled ones from J.

Disposable and short-life articles

Although articles in the linen room have been described as launderable, it should be remembered that much more use is now made of plastic and disposable materials which may replace linen. Great advances have been made in the manufacture of plastic tablecloths, mats and traycloths. These are durable, economical in price and are not sent to the laundry; when soiled they are sponged or washed on the premises.

Disposable articles are generally those thought of as being used once and then thrown away, eg paper napkins, tissues and hand towels, and from this point of view they are very hygienic.

There are 'disposables' made of rayon which may be used more than once and yet have a short life compared with their normal counterpart, eg bed linen and cleaning cloths. In the case of cleaning cloths which depend on use and user, articles may be used once or may withstand washing and further use, whereas bed linen may stand up to several nights' use and will then be disposed of.

These articles are hygienic and labour saving, but problems which must be overcome are the collection and disposal of the soiled articles and storage space for the new stock. A stock of disposable bed linen is useful in case of emergency, eg strikes, and is more likely to be found in hostels and hospitals than in hotels.

Quantity and control of linen

The quantity of table and bed linen required by an establishment will vary considerably, depending on the type of trade carried on. As far as bed linen is concerned, one factor to be taken into account is the frequency with which the beds are changed. A luxury hotel will re-sheet each day, other hotels will re-sheet every second or third day, while yet other establishments only once a week, and this also applies to staff quarters. In all establishments, re-sheeting is always done after a departure and this may mean that where there are many 'one nighters', many beds will in fact be re-sheeted each day.

Another factor which will influence the amount of linen is the frequency with which the laundry collects and delivers. A good commercial laundry in a town normally takes 48 hours between collection and return of the articles and for large hotels will pick up and deliver each day but this does not hold good over weekends and public holidays.

For one bed there could be one pair of sheets in use, one soiled pair waiting to go to the laundry, one pair at the laundry and one pair in the linen room or in the floor linen cupboard. So for many hotels where re-sheeting is done approximately every third day (and remembering there may be some 'one nighters'), four pairs of sheets per bed should be sufficient, but this takes no account of articles put aside for re-washing and repairing, and for non-delivery of laundry at weekends and public holidays. The same thinking can be applied to pillowslips and towels, remembering that double beds have two pairs of pillowslips, and that in some hotels towels are changed more frequently than bed linen, and so more towels would be needed than sheets or pillow slips. In luxury hotels, because of the daily re-sheeting, five to eight pairs of sheets per bed is more usual. Where the complete re-sheeting is once a fortnight, ie one sheet and one slip per week as in many hostels, the number of sheets required is two to three pairs. Similar calculations can be made for restaurant linen on the basis of number of covers, size of tables and number of sittings.

Table cloths, table napkins, waiters' cloths, glass cloths, tea towels and oven cloths are issued from the linen room as the restaurant and kitchen require. The appearance of the cloth on the table is of great importance and linen cloths are crisper, smoother and have a natural sheen and better defined folds than cotton ones, but they are of course much more expensive. Linen/polyester and cotton/polyester blends are also used for cloths. A patterned weave is generally considered more attractive than a plain weave, so cloths are traditionally woven in the damask weave with its self-coloured pattern (see page 212). (For bed linen, see pages 234–7.)

Even with a small establishment many hundreds of articles will be necessary and these represent a large sum of money. All too frequently, too little thought is given to their cost when considering the price of a bedroom or meal. In both cases there is the initial cost, laundering, care and replacement costs of the articles to be remembered and without proper control these may be excessive.

Control of the linen starts in the linen room where first steps are taken to prevent loss and damage, and this control should continue through every department handling linen. There will inevitably come a time when replacements are necessary due to normal wear and tear, excessive wear and tear and loss, and the last two may be due to lack of control. Training and supervision of the staff is essential.

Lack of control may lead to:

misuse of linen by waiters and maids;

insufficient care of damp and stained linen, resulting in mildew and the spread of iron mould;

carelessness in stripping beds, resulting in sheets getting torn by castors;

lack of adequate protection during storage, resulting in articles becoming marked and the need for extra laundering;

lack of inspection, resulting in torn articles being used and tears becoming worse;

insufficient stock and poor rotation, resulting in linen not resting between laundering and the next use;

careless handling, resulting in soiling, creasing, etc;

cupboards and linen room being left unlocked, resulting in loss of linen.

Buying linen

It is an economy to buy good quality material for any establishment because of the great use and the frequent launderings to which the linen is subjected. It is advisable to get prices and see samples from more than one source, and weigh up the merits of each against the cost. The largest quantity possible should be ordered at one time in order to get the cheapest rates, and so that exact requirements may be met orders should be placed early. (See page 93.)

Samples may be tested by:

a) rubbing the material between the hands over dark material and noting the amount of dressing, ie starch, which falls on to it; if much falls it denotes a poor quality material;

b) looking at the material under a magnifying glass to note the closeness and evenness of the weave;

c) noting the firmness of the selvedge and the finish of the machining, especially at the corners;

d) sending a sample of the material to the laundry to be washed a given number of times, and comparing it with a once-washed sample, to get some idea of the wearing quality.

Samples may also be sent to testing houses to get an idea as to fibre content and count.

There are distinct advantages in an hotel owning its own linen in that it has the full choice of quality, size and colour of the articles and it may have its own monogram. However, a large sum of money has to be found initially and whenever replacements are needed. Mending and laundering have to be arranged and these may be done on or off the site.

Firms will undertake the marking of articles at a small extra cost. Where large numbers of good quality table linen and towels are ordered, the name, initials or crest may be woven into the fabric. In other cases embroidery may be used. Embroidered names may be worked on most articles, except such expendable items as dusters and kitchen cloths. *Note* The weaving and embroidering of the names undertaken by the manufacturers must not be confused with machine marking done in the linen room.

Linen hire

Owing to the high cost of linen and its upkeep, the hiring of linen from firms offering a linen rental service has become more popular. The firms undertake to supply clean articles in good condition and arrangements are made between the firm and the house regarding the amount of linen required, the frequency of deliveries and the price to be charged. Stocktaking is still normal practice and losses have to be paid for.

The advantages are:

it cuts out the heavy initial cost of buying linen;

no large sum of money is required for replacements; it cuts out the need to order new linen;

the cost of hiring, which includes laundering, comes from revenue;

linen hire charges may be no greater than the combined depreciation and laundering costs;

short term loans are possible for special occasions, eg banqueting and so there may be less stock and less storage space;

no repairing of linen on the premises is necessary;

less space is required for the linen room;

fewer staff are necessary and therefore there are fewer wages to pay;

various sizes of such things as overalls, waiters' jackets and chefs' uniforms are available;

in some instances, and if stock is returned, there will be no cost to the establishment which closes for part (or parts) of the year.

The disadvantages are:

little choice regarding quality and style;

quality is variable;

standards are not always maintained;

no rags are available from the linen room;

no renovated articles, eg cot sheets, under pillowslips etc, are available;

the contract price remains the same even when numbers fall over a short period (ie there is normally a minimum charge).

10

LAUNDRY, DRY CLEANING
AND STAIN REMOVAL

The cost of linen and its laundering is so high that it is sensible for anyone responsible for linen to know a little of the work done in a laundry. It is an advantage if the *housekeeper* and/or the linen keeper visits the laundry, so that misunderstandings may be prevented and good co-operation ensured.

In an establishment where linen is changed very frequently, it is more 'washed out' than 'worn out' and so the life of a cotton sheet, for example, is often given as perhaps 200–250 washes rather than four to five years. But (and it is a big but) the 'life' depends on the care of the linen in use and the treatment it gets at the laundry; not forgetting that good quality material needs to be bought in the first instance.

A good laundry is therefore of great importance to any establishment in order that:

articles are handled carefully,
tensile strength of the material is not impaired,
white material is kept white,
stains are removed when requested,
materials are not ruined by excessive use of bleach,
lists are checked carefully so that there are few 'shorts',
the work is carried out as speedily as possible,
good co-operation is maintained regarding damage and losses.

An establishment has the choice of the following laundry services:

- commercial,
- in-house or on premises,
- linen hire, which includes laundry service.

None of these can be automatically considered as offering the best and most economical laundry service; each has advantages and disadvantages.

Commercial laundry

At a commercial laundry the following work is carried out on white cotton and linen articles.

At the laundry the linen is checked and sorted into groups, then a suitable number or weight of similar articles, eg sheets, are put into a:

washing machine – a large drum which revolves first in one direction and then in the other to prevent tangling of the articles. Agitation, hot water at 85°–94°C and suitable detergents (usually synthetic with a small quantity of soap added to act as a lubricant) bring about the cleansing action, after which rinsing in several waters takes place and all these processes may be automatically controlled. The more sophisticated equipment is controlled by computer.

Most laundries soften their water and often a 'brightener' is added to help keep whites white. Heavily soiled articles, eg kitchen rubbers, may need a little bleach in the washing water, but it should only be used when really necessary as it weakens the material. Table linen may need a little starch and this is added to the last rinsing water.

The clean articles are then passed into a:

hydro-extractor – which whirls the water out of the articles leaving them as a very tightly packed mass which needs shaking out in a special tumbler.

However, there are modern washer extractors with a dry weight capacity of about 5–400 kg which can wash, rinse, hydro-extract and shake out all in the one machine.

The articles may now be put through a:

calender or *ironing machine* – this very large machine consists of several heated and well padded rollers which iron the article as it passes through (see Fig. 10.1). Only flat articles are calendered and a large calender will be wide enough to take a double sheet. After ironing, the articles are folded either by an automatic electric device or by hand and they are then ready for sorting, packing and sending back to the establishment. Turkish towels are not normally ironed nowadays but tumble dried and folded. Strictly speaking, table cloths and napkins should be folded in a screenfold of three; bedlinen and towels should be folded lengthwise.

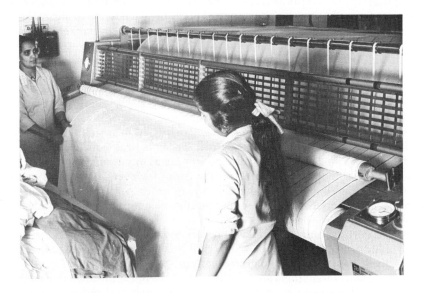

Figure 10.1 Calender or ironing machine

A 50/50 blend of cotton and polyester is now popular for sheets and pillowslips but, because the resin finish is harmed above 80°C, these should be washed at a lower temperature than cotton articles. Ironing is unnecessary if articles are tumble dried and folded while still warm, but this is not possible at a commercial laundry.

Shaped articles, such as shirts, nurses' uniforms, white coats, waiters' jackets, overalls, etc, are not ironed through a calender but are dealt with on presses and/or by steam inflated 'suzies'. A 'suzie' is an inflatable shape onto which, for example, a shirt is placed and which is then filled with steam, so drying and pressing out the creases of the shirt.

In a *hospital laundry* a section of the 'soiled' side should be used as the 'central disinfection area' into which the waterproof bags containing fouled and/or infected linen are delivered. The contents of the bags are tipped straight into double-ended washing machines (tunnel washers) sited between the 'central disinfection area' and the 'clean side'. The fouled and/or infected linen should be sluiced (if fouled) and washed in these machines using the times and temperatures recommended to achieve disinfection (ie 70°C for 3 minutes).

Where a laundry has no central disinfection area, arrangements are made for the fouled and/or infected linen to be sluiced, disinfected and dried at the hospital before being sent to the laundry.

Any fouled and/or infected linen in an hotel, hostel etc, should be sluiced, disinfected and laundered on site.

Blankets are possibly the only woollen articles which an establishment sends to the laundry (they may, of course, be dry cleaned) and here the problem is keeping them fluffy. To do this, blankets should be washed in cool water (40°C) with a suitably mild detergent, a minimum amount of friction and avoiding the use of soda and bleach. They are dried on racks or put into a heated machine which tumbles them dry. Acrilan blankets may be laundered more satisfactorily than woollen ones; there is little tendency to shrink and their finish after tumble drying is excellent. Cotton cellular blankets as used in hospitals have a fire retardant finish and should be marked so that they receive the correct laundering procedure. Other articles treated are cubicle curtains and primary and secondary duvet covers.

Articles with loose colour, ie colours which run, such as yellow dusters, must be kept separate from other articles during the washing and drying process.

Nylon articles are very easily washed and dried and so are not often sent to the laundry, but when they are the finish presents problems. Fitted nylon sheets when washed at home require no ironing as any rough appearance is soon lost when they are stretched on the bed. However, a laundry has to return articles with a good appearance and in order to do this nylon sheets need special care in ironing, consequently they cost more to have laundered than cotton or linen sheets.

Unlined curtains and bedspreads are probably the main rayon articles sent to the laundry and these need care in washing as rayon fibres are weak when wet and will not stand up to high temperatures.

Fresh tea and coffee stains are usually removed from white linen during the ordinary washing process, but other stains, eg rust and mildew, require special treatment and this is not undertaken by the laundry unless specially requested, when a charge will be made. Stained articles for special treatment should therefore be sent to the laundry in a separate container.

Some large establishments have contracts with laundries, whereby a price is agreed by number or weight of flat articles provided a minimum number is sent.

In-house laundry

There is a growing tendency for establishments to have their own laundry on the premises. The reasons for this may be that:

with the advent of polyester/cotton materials the use of a large expensive calender is no longer required, so laundry premises can be smaller and the initial outlay on equipment less;

there is greater variety in size of laundry equipment available, resulting in full use of the equipment chosen;

articles in demand can be dealt with out of turn and under normal circumstances there is a quicker turnround and so less stock is required;

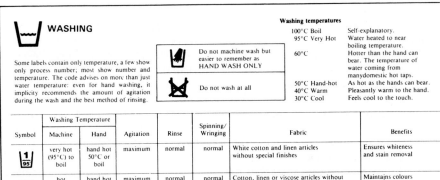

Symbol	Washing Temperature		Agitation	Rinse	Spinning/ Wringing	Fabric	Benefits
	Machine	Hand					
$\frac{1}{95}$	very hot (95°C) to boil	hand hot 50°C or boil	maximum	normal	normal	White cotton and linen articles without special finishes	Ensures whiteness and stain removal
$\frac{2}{60}$	hot 60°C	hand hot 50°C	maximum	normal	normal	Cotton, linen or viscose articles without special finishes where colours are fast at 60°C	Maintains colours
$\frac{3}{60}$	hot 60°C	hand hot 50°C	medium	cold	short spin or drip dry	White nylon; white polyester/cotton mixtures	Prolongs whiteness —minimises creasing
$\frac{4}{50}$	hand hot 50°C	hand hot 50°C	medium	cold	short spin or drip dry	Coloured nylon; polyester; cotton and viscose articles with special finishes; acrylic/cotton mixtures; coloured polyester/cotton mixtures	Safeguards colour and finish — minimises creasing
$\frac{5}{40}$	warm 40°C	warm 40°C	medium	normal	normal	Cotton, linen or viscose articles where colours are fast at 40°C, but not at 60°C	Safeguards the colour fastness
$\frac{6}{40}$	warm 40°C	warm 40°C	minimum	cold	short spin	Acrylics; acetate and triacetate, including mixtures with wool; polyester/wool blends	Preserves colour and shape — minimises creasing
$\frac{7}{40}$	warm 40°C	warm 40°C	minimum do not rub	normal	normal spin do not hand wring	Wool, including blankets, and wool mixtures with cotton or viscose; silk	Keeps colour, size and handle
$\frac{8}{30}$	cool 30°C	cool 30°C	minimum	cold	short spin do not hand wring	Silk and printed acetate fabrics with colours not fast at 40°C	Prevents colour loss
$\frac{9}{95}$	very hot (95°C) to boil	hand hot 50°C or boil	maximum	cold	drip dry	Cotton articles with special finishes capable of being boiled but requiring drip drying	Prolongs whiteness, retains special crease resistant finish

WASHING

Some labels contain only temperature, a few show only process number; most show number and temperature. The code advises on more than just water temperature: even for hand washing, it implicitly recommends the amount of agitation during the wash and the best method of rinsing.

Do not machine wash but easier to remember as HAND WASH ONLY

Do not wash at all

Washing temperatures

100°C Boil	Self-explanatory.
95°C Very Hot	Water heated to near boiling temperature.
60°C	Hotter than the hand can bear. The temperature of water coming from many domestic hot taps.
50°C Hand-hot	As hot as the hands can bear.
40°C Warm	Pleasantly warm to the hand.
30°C Cool	Feels cool to the touch.

Figure 10.2 Laundering labels

it may be possible to rely on staff to inspect the linen, so saving work in the linen room;

there is more freedom in laundering methods used and the possibility of a greater life expectancy of the linen or other article;

there is internal supervision and security, which may result in fewer losses;

there are no transport difficulties and costs.

It is possible to hire laundry equipment instead of purchasing which saves initial outlay but as in all cases of hiring it needs careful consideration.

For a number of years, of course, some places have used domestic laundry equipment for such items as towels and non-iron articles (see Fig. 10.2 for details of laundering care). There are also halls of residence, hostels and hotels where laundry facilities are provided for guests.

The third laundry service to an establishment is linen hire which includes the laundering and this is dealt with on page 133.

Dry cleaning

The responsibility for the sorting, dispatching, receiving and storing of hotel articles to be dry cleaned may be that of the linen keeper or an assistant housekeeper.

In addition to the dry cleaning of blankets, curtains, quilts, etc, some staff have a dry cleaning allowance for their 'working dress'. In an hotel, guests' dry cleaning is dispatched by the valet, hall porter or the linen keeper. Articles for dry cleaning may be sent to a firm of dry cleaners or to a laundry with a dry cleaning department.

On arrival at the cleaners, each article is marked with an identifying tape, checked for special stains and items in the pockets and brushed free of loose dust. Certain types of stains, such as blood, egg or sugary substances etc, are removed more easily if some attention is given to them before cleaning, so these stain areas are 'pre-spotted'.

Although in certain instances different coloured articles may be cleaned successfully together, it is normal to divide the work into a number of classifications, such as whites, mediums, darks. This ensures that heavily soiled articles are not cleaned with lighter soiled ones and dark coloured lint does not then transfer to light coloured articles or vice versa.

The prepared articles are then 'washed' in a dry cleaning solvent, such as perchlorethylene, in an enclosed machine in which the 'washing', extraction and drying are all carried out in the same cage. The solvent, because of its cost, is not wasted but distilled and/or filtered for re-use.

Dry cleaning solvents do not affect textile fibres in the same way as water, so when dry cleaning some materials the risk of shrinkage, severe creasing, distortion or colour movement may be greatly reduced. (*Note* Shrinkage can still occur, partly because of the agitation and partly because a small amount of water may be present.)

Because water-borne soil and stains can only effectively be removed by the use of water, a controlled amount of water and detergent is introduced into the solvent during some dry cleaning processes (known as 'charged system'). After the articles have been cleaned, they are spun dried to extract the bulk of

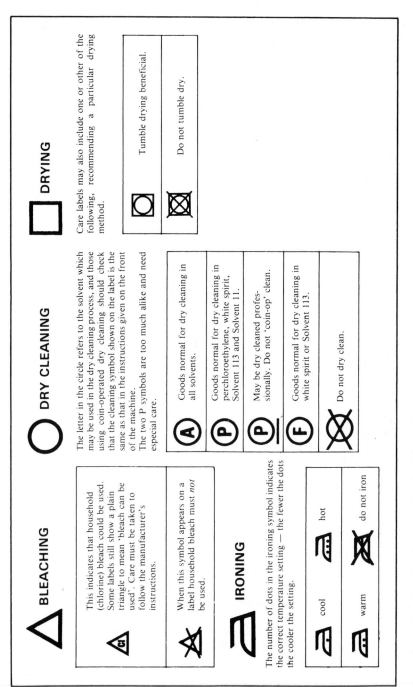

BLEACHING

This indicates that household (chlorine) bleach could be used. Some labels still show a plain triangle to mean 'bleach can be used'. Care must be taken to follow the manufacturer's instructions.

When this symbol appears on a label household bleach must *not* be used.

IRONING

The number of dots in the ironing symbol indicates the correct temperature setting — the fewer the dots the cooler the setting.

cool	hot
warm	do not iron

DRY CLEANING

The letter in the circle refers to the solvent which may be used in the dry cleaning process, and those using coin-operated dry cleaning should check that the cleaning symbol shown on the label is the same as that in the instructions given on the front of the machine.
The two P symbols are too much alike and need especial care.

Ⓐ	Goods normal for dry cleaning in all solvents.
Ⓟ	Goods normal for dry cleaning in perchloroethylene, white spirit, Solvent 113 and Solvent 11.
Ⓟ	May be dry cleaned professionally. Do not 'coin-op' clean.
Ⓕ	Goods normal for dry cleaning in white spirit or Solvent 113.
⊗	Do not dry clean.

DRYING

Care labels may also include one or other of the following, recommending a particular drying method.

⊡	Tumble drying beneficial.
⊠	Do not tumble dry.

Figure 10.3 Bleaching, dry cleaning and drying information

the solvent and then dried with warm air. The cleaned articles are hung up to remove the smell and checked for stains, and any remaining stains are dealt with before pressing. If any special finish is required, eg re-texturing, this is also done before pressing and the fabric is treated with a solvent containing resin to improve body and bulk.

After pressing, the articles are returned in hampers, boxes or on coat hangers and, in the case of guests' dry cleaning, the cost is entered on the guest's bill.

Stain removal

Many fresh stains, eg tea, coffee, grease etc, will be removed from cotton and linen articles during the normal washing process. Protein stains, eg egg, blood, glue, perspiration etc, are more easily removed by pre-soaking in lukewarm water with a detergent containing enzymes which digest the protein. (*Note* Enzymes are inactive in hot water above 40°–50°C.) All stains should be dealt with as soon as they occur or as soon after as possible.

Stain removal agents

If old or heavy, stains require special treatment with stain removal agents. The use of these stain removal agents requires care as they can cause weakening of the fibres, bleeding of dyes, damage to special fabric finishes, and some are inflammable while others are poisonous. There are five main stain removal agents: organic solvents, acids, alkalis, bleaches and enzymes.

Organic solvents

For example:

(a) benzene	(b) carbon tetrachloride
acetone	perchlorethylene
amyl acetate	trichlorethylene
methylated spirit	
white spirit (turpentine substitute)	

These dissolve grease and require care in use because:

group (a) is inflammable and should never be used near a naked flame;
group (b) is non-inflammable but harmful when inhaled and should be used only in a well-ventilated area.

In the main, fibres and dyes are not harmed by these solvents but acetone dissolves rayon acetate, trichlorethylene harms triacetate, spirit affects rubber (rubber-backed carpets) etc. When using a solvent always work from the outside of the stain inwards, with an absorbent cloth underneath the fabric.

Chewing gum (after scraping), grease, oil paint, lipstick, ball-point ink, etc, will sometimes yield to a solvent.

Acids

Acids include oxalic acid, potassium acid oxalate (salts of lemon), and various rust removers sold under trade names. (All these are poisonous.) Fibres vary in their susceptibility to damage by acids. Dilute acids can be used on most white fabrics but many coloureds are affected by acids. It is always better to use a weak solution several times than use a stronger solution at first. After treatment, washing using a detergent or thorough rinsing in a weak alkaline solution is essential to neutralize the acid and to prevent damage to the fabric (the acid concentrates on drying), always remembering that alkalis affect animal fibres.

Acids remove metal stains, the commonest of which are iron mould or rust and the iron stain left after washing a blood-stained article.

Alkalis

Alkalis such as soda and borax, remove old and heavy vegetable stains, eg tea, coffee, wine etc, from white linen or cotton fabrics. Animal fibres and dyes may be adversely affected.

Bleaches

The process of changing a coloured substance into a colourless one is known as bleaching, ie bleaches whiten. Bleaches also weaken fibres so extreme care is needed in their use.

Bleaches are of two types:

oxidizing	*reducing*
eg sodium hypochlorite	eg sodium hydrosulphite
hydrogen peroxide	
sodium perborate	

Oxidizing bleaches liberate oxygen from themselves or other substances. The most frequently used oxidizing bleaches are those named above.

Sodium hypochlorite (normal household bleach) damages animal fibres and so should not be used on woollen or silk articles. It is used mainly for the removal of obstinate stains on cotton and linen fabrics but it 'fixes' iron stains. All fabrics should be thoroughly rinsed after being treated with hypochlorites or the fabrics will rot. An added hazard with 50/50 polyester/cotton fabrics is that the resin tends to retain the chlorine. In commercial use, an anti-chlor, eg sodium thiosulphate ('hypo'), is added to the final rinse to remove all traces of the free chlorine.

Hydrogen peroxide is slower acting than the hypochlorite bleaches and can be used on most white fabrics. The peroxide decomposes more readily if the solution is rendered just alkaline with ammonia.

Sodium perborate is the bleach present in powdered soap and soapless detergents. It is safe to use on most fabrics and is most effective at temperatures above 85°C.

Reducing bleaches remove oxygen or add hydrogen to the coloured substance, sodium hydrosulphite is the most frequently used. It can be used on most white fabrics and is used for the removal of iron stains and the stripping of dyes. It is in general milder in its action than the oxidizing bleaches. After bleaching by reduction there is a tendency for white articles, eg white woollen blankets, to take up oxygen – particularly in sunlight – and become yellowed.

Enzymes

Enzymes, eg powdered pepsin, work most effectively at a temperature of 40–50°C and are used to remove protein stains, eg blood, perspiration, egg etc, from all fabrics.

When the origin of a stain is known the specific stain removal agent can be used straight away, but if unknown it may be necessary to try several agents before the right one is found. In general, safer treatments are tried first and it is better to repeat a process twice with a weak solution than to use a strong solution at the beginning. It is essential that the agents are completely removed from the fabric by evaporation or neutralization, washing or thorough rinsing.

To treat an unknown stain:

1 Soak in cold water
2 Dry and use a grease solvent
3 Use an acid
4 Use an alkali

Stains on coloured materials are very difficult to remove as many of the stain removal agents affect dyes. In the case of carpets and upholstery, stains are particularly difficult to remove because they have to be dealt with *in situ* and the colour, the backing and the padding may present problems. Grease absorbers in the form of aerosol sprays may prove useful.

It must be strongly emphasized that owing to the variety of fibres used in modern materials and the unknown qualities of some stains, stain removal is a highly skilled job and should not be undertaken lightly.

Stain repellants

Fabrics may be treated so as to be made 'stain repellant' and this may be achieved by the use of fluorochemicals, eg Scotchguard, which will give both water and oil repellancy. The stains tend to stand on the surface and can be blotted away (not wiped). These finishes are expensive but there is no change in colour or texture of the fabric, and they withstand dry cleaning and at least five washes.

More usually, the fabrics are made water repellant only, by the use of silicones, eg Velan, Drisil, when water-borne stains will not wet the surface and so can be blotted away. The fabric will still absorb oil-borne stains but even they are more easily removed with solvent cleansers which do not remove the silicone finish.

Polyurethane is sometimes used as a very thin flexible coating on some fabrics intended to be waterproof.

Specific stains

For the more usual stains on white and fast-coloured fabrics; the following stain-removal agents are suggested:

Ball-point ink: methylated spirit or carbon tetrachloride.

Blood – new: soak in cool or warm detergent solution.

 – old: treat as iron mould.

Chewing gum: rub with ice-cube and scrape.

Dyes: bleach (not chlorine bleaches on animal fibres).

Grass: eucalyptus oil or glycerine, follow with spirit or washing.

Ink: if not removed by washing treat as for iron mould.

Ink (red): often not removable, except when very fresh, but some may respond to washing or sodium hydrosulphite.

Iron mould: rust remover, oxalic acid, potassium acid oxalate (salts of lemon), sodium hydrosulphite or Rustasol.

Lacquer and nail varnish: amyl acetate, acetone (not on rayon acetate) or a cellulose thinner.

Lipstick: carbon tetrachloride and/or sodium hydrosulphite.

Mildew: hot weak potassium permanganate solution followed by a weak acid or hydrogen peroxide.

Paint (oil): if fresh, white spirit, or a proprietary paint remover followed by a solvent.

Paint (cellulose): amyl acetate, acetone (not on rayon acetate) or a commercial cellulose thinner.

Paint (emulsion): wash immediately, as once dried it is almost irremovable.

Perspiration: treat as for mildew or protein stains.

Protein stains, eg egg, meat, perspiration: protein digesting enzyme contained in biological detergents or as powdered pepsin.

Tar: carbon tetrachloride or white spirit, scraping first.

Vegetable stains, eg tea, coffee, etc: alkali or bleach (not chlorine bleaches on animal fibres).

Vomit: scrape, soak and wash. If not washable, sponge with warm water containing a few drops of ammonia. Blot dry.

Stain removal from different surfaces

Carpets and upholstery (care must be taken not to wet the backing or padding)

Candle grease: scrape, use hot iron and absorbent paper. Follow if necessary with a grease solvent.

Ink: mop up as quickly as possible to prevent spreading. Wash with warm water and synthetic detergent or use a weak acid, and rinse.

Mud: leave to dry, then brush off.

Shoe polish: scrape off if possible and then apply a grease solvent.

Urine: sponge with salt water, followed by a weak solution of ammonia and rinse well or a squirt from a soda water siphon.

Polished wood

Ink: mop up as quickly as possible. Rub with fine dry steel wool or glass paper, or dab with a hot solution of weak acid and rinse. In both cases colour and polish will be removed, so rub with linseed oil or shoe polish to darken and later apply polish, and buff well.

Spills, slight heat and burn marks: rub with a rag moistened with a drop or two of liquid metal polish or methylated spirit and then repolish, or rub with a very fine abrasive, eg cigarette ash or very fine steel wool and repolish.

Scratch marks: if newly scratched cover with iodine, potassium permanganate solution or shoe polish according to the colour of the wood. If necessary remove polish first with a mild abrasive.

Alcohol: (a) wipe up, rub with finger dipped in silver polish, linseed oil or cigarette ash. Repolish.

(b) wipe up, put *few* drops of ammonia on damp cloth and rub. Immediately repolish.

Wood with oil finish

Small burns and heat marks: rub with emery cloth or fine sandpaper, followed by boiled linseed oil.

Marble, terrazzo

Ink: apply a poultice of sodium perborate, precipitated whiting and water. Leave to dry.

Rust: apply a poultice of sodium citrate crystals, glycerine, precipitated whiting and water. Leave to dry.

Points to remember

When removing stains, it is worth remembering the following:

1 Treat stains as soon as possible.
·2 Consider the fibres of which the fabric is made.
3 If a coloured article, check effect of remover on an unimportant part if possible.
4 Use the weakest methods first.
5 Use a weak solution several times, rather than one strong one.
6 When using a chemical always place the stained area over an absorbent pad of clean cloth.
7 To avoid a 'ring' always treat from an area round the stain and work towards the centre.
8 After using a chemical, neutralize or rinse well.

11

PESTS AND WASTE DISPOSAL

Pests

Moths

Clothes and house moths are of a pale buff colour and are seen flying mainly between June and October. They are relatively small and rarely live for longer than a month.

The female lays its eggs (up to 200) in some dark, warm place on material which the grubs (larvae) will later eat. The eggs hatch and the grubs immediately feed on the material as they move about. When fully grown they crawl into sheltered places, spin a cocoon round themselves, become a chrysalis (pupa) and later emerge as moths to start another life cycle. The entire life cycle (egg – grub – chrysalis – moth) varies from one month to two years depending on the food available, temperature and humidity.

The materials which are attacked by moth (the grubs) are wool, fur, skin and feathers, and those which are immune are rubber, man-made and vegetable fibres. Thus it follows that the articles which most need protection from damage by moth are:

blankets, bedding and quilts (not man-made fibres);
carpets and underfelts;
upholstered furniture and curtains;
stuffed animals and birds, ie fur and feathers.

While feeding on these materials the grubs form small holes in the articles and damage occurs frequently during storage, because of the warmth, darkness and lack of disturbance. It is advisable that articles to be stored should be clean (ie vacuum cleaned, brushed, washed or dry cleaned), protected by a moth deterrent and inspected frequently. Commonly used moth deterrents are naphthalene, camphor tablets and paradichlorbenzene, whilst insecticides containing pyrethrum are used to kill the pests.

Upholstered furniture and the edges of carpets are often attacked by moth, and thus brushing and/or vacuum cleaning with a suitable attachment must be done at frequent intervals. Materials may be treated by a chemical process to render them immune from moth (see page 216).

Where moth has attacked, for example a blanket or rolled up piece of carpet, and the articles are found full of live grubs, the safest thing is to burn them. Temperatures of 60°C and above will destroy grubs and eggs and infested articles such as upholstered furniture may be treated by heat, providing no harm will come to the articles, or they may be fumigated, and in both cases the treatment should be carried out by experts.

Carpet beetles

Carpet beetles are 2–4 mm long like small mottled brown, grey and cream ladybirds. The larvae are small, covered in brown hairs and tend to roll up when disturbed. As they grow, they moult and the old cast-off skins may be the first sign of infestation. Adults are often seen April–June, seeking places to lay their eggs and the larvae are most active in October before they hibernate.

The adult beetle feeds on pollen and nectar of flowers but lays its eggs in old birds' nests, fabrics and accumulated fluff in buildings. It is the larvae which hatch from these eggs that do the damage by feeding on feathers, fur, hair or wool and articles made from these substances. They tend to wander along pipes from the roof to storage cupboards and under floorboards to carpets and underfelts. Carpet beetles are now the major textile pest and do more damage than moths. The damage consists of fairly well-defined round holes along the seams of fabrics where the larvae have bitten through the threads.

Frequent vacuum cleaning of fluff and debris from storage cupboards, floorboards, carpets and upholstery is the main means of control. Insect killing powders may be sprayed between floorboards, under carpets and underfelts and in crevices where fluff may collect and attract the larvae.

The life cycle takes about a year and the larvae can survive for several months without food.

Wood-boring beetles

These beetles can be likened to moths or carpet beetles, in that it is the grub, larva or 'worm' which does the damage to the wood. The common furniture beetle lays about 20–60 eggs in cracks and crevices of unpolished wood, eg flooring, panels, roof timbers, backs of wardrobes and chests of drawers. On hatching, the grub eats its way through the wood, and this tunnelling causes weakening of the wood and may take from 2–3 years.

Eventually the grub matures, bores towards the surface of the wood and changes into a pupa. From this pupa emerges the beetle, which bites its way into the open air through an exit hole which is about 0.15 cm in size. The beetles have a very short life (probably 2–3 weeks) during which time they move around by walking or occasionally flying, mating takes place and eggs are laid, often in the old exit holes.

Many pieces of wood have exit holes in them but they may have been successfully treated and are therefore inactive. If, however, small piles of bore-dust are found beneath the holes, these indicate the presence of active 'worms' in the wood and treatment is necessary; wood which is infested in time has innumerable tunnels in it and its strength is badly impaired.

Eggs are laid on unpolished wooden surfaces, so the use of shellac, varnish,

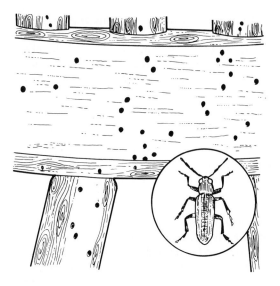

Figure 11.1   Exit holes of the common furniture beetle

lacquer or polish acts as a deterrent. To kill woodworm, the exit holes should be sprayed, brushed or injected several times with one of the proprietary oil soluble woodworm fluids, eg Rentokil or Cuprinol. This may be done by the individual, or experts may be called in to do the job more thoroughly. There are other treatments, such as heat and fumigation with poisonous gases, but these have disadvantages, do not prevent re-infestation and their use is best left to experts. A badly infested piece of wood is better burnt.

Fleas and other insect pests

These insect pests are frequently, though not necessarily, associated with dirt. They abound in unhygienic conditions and their entry into clean places may be entirely accidental.

There are many different kinds of *fleas* and each has a preference for one kind of host, eg human, cat, dog, vermin, and any of these hosts may introduce fleas into an establishment. Fleas bite their hosts, causing annoyance, and in the human, large red, itching spots appear on the skin.

Fleas are able to cover considerable distances because of their jumping powers; they like darkness and warmth and are capable of laying a large number of eggs, generally in the cracks of floors. Spraying with insecticides is a suitable way of eradicating them.

Head lice, which live in the hair of the head, are probably the most common of all lice. They cause intense irritation and suck blood; the eggs, 'nits', which are very numerous, are stuck firmly on to the hairs and cannot be removed by brushing. Lice may be caught from upholstery in trains, by borrowing combs, etc. Frequent brushing, fine combing and when necessary, the use of a special shampoo are the best remedies.

Bedbugs may be secreted in second-hand furniture, bedding and books and under the wooden lathes of trunks, when luggage has been lying in the holds of ships or in trains, and thus may find their way into any establishment.

Bedbugs are about 5 mm long and 3 mm wide, reddish brown in colour, and are able to survive sometimes many months without food. They are nocturnal by habit, feed by sucking human blood and deposit their eggs in cracks and crevices of woodwork, behind wallpaper, etc. The eggs are stuck to these surfaces by a cement-like substance exuded by the bug and are therefore difficult to brush off.

Bedbugs cover considerable distances although they cannot fly and they give off a very unpleasant smell. Their bites cause considerable irritation and may result in large red patches, with swelling on some people, and the reporting of this condition may well be the first indication that there are bugs in the establishment.

Bugs may be exterminated by spraying with a suitable liquid insecticide, by heat treatment or by fumigation, which is normally carried out by experts. It is important that instructions are followed carefully as some chemicals cause damage to metals and coloured fabrics, and heat may loosen joints of furniture. It must, however, be emphasized that if articles are removed from the room, they should be thoroughly inspected first.

Silverfish are wingless insects, silvery grey in colour and about 1 cm long. The young closely resemble the adults and both are rounded in front and tapered towards the rear. Silverfish require a moist place in which to live and are found in basements, and around pipes, drains, sinks, etc. They leave their hiding places in search of food of a cellulose nature. They feed on starchy foods, paste in wallpaper and books, and may attack clothing made of cotton or rayon, especially if starched. They may be prevented by regular cleaning of cupboards and surroundings to sinks, pipes, etc, and insecticidal powders can be sprinkled where silverfish have been seen.

Cockroaches are more likely to be found in the kitchen and restaurant/dining room areas than in the accommodation area, although cockroaches do not necessarily require human food, and will feed on whitewash, hair and books if no other food is available. Hygienic storage and disposal of food and waste and the cleanliness of all areas where food is handled are important points in the prevention of an infestation.

Cockroaches are difficult to eradicate but a residual insecticide, eg chlordecone in a raw liver base, may be used in cracks and crevices and holes, especially in brick or plaster work through which warm pipes pass. The Public Health Department will help if the problem persists.

Flies are of many types and most are attracted to food and food waste. Aerosol sprays are useful in ridding an area of the many varieties that may gain entry to a building.

Rats and mice

Rats and mice are more likely to be found in kitchens and dining areas than in bedrooms. They are attracted by scraps of food, candles, soap etc. Hygienic storage and disposal of food and all kinds of waste, and the cleanliness of all areas where food is handled, are important to prevent an infestation. Rats and mice may be destroyed with poisons. Rats and bad infestations of mice may also be dealt with by experts from the local Public Health Department.

Wood rot

Dry rot

This is the term used for the decay of timber by a fungus which grows and lives on the wood, and reduces it finally to a dry crumbling state – hence the name 'dry rot'. It nearly always starts in damp (more than 20 per cent moisture) unventilated places, behind wood panelling, under floorboards, and spreads by sending out thin root-like strands which creep over brickwork to attack surrounding wood. Once a fungus gets a hold, it produces 'fruit bodies' which are flat pancake-like growths with spore-bearing centres. The spores are produced in enormous numbers and are so small that they appear as reddish brown dust which may be blown about very easily and for great distances.

Dry rot can be recognized by its offensive, mouldy smell, by its friable condition and the 'dead' sound when the wood is hit with a hammer. When dry rot occurs it is necessary to find the reason for the dampness of the wood, and the following are some of the more common causes:

slightly leaky joints behind the bath panels, or any faulty plumbing keeping the floorboards damp;
not drying out wet boards under a floor covering such as linoleum;
no damp proof course;
ineffective damp proof course through the piling up of earth, coal, sand, etc against the outside wall;
broken damp proof course.

Having ascertained and cured the cause of the dampness, all rotten wood must be cut away 30–60 cm beyond the infected area and burnt at once – never stored. All brickwork which was near the infected wood should be sterilized by the use of a blow lamp and, when cool, treated with a preservative before repairing with dry, well seasoned timber which should also have been treated with a preservative.

Wet rot

This is the name given to the fungal decay in timber in very damp situations. The fungus usually involved is the cellar fungus and it attacks timber that is definitely wet. In view of this, it is commonly found in cellars, roofs and bathrooms, in fact in any place where leakage of water is liable to occur. It requires considerably more moisture for development than the dry rot fungus, the optimum moisture content being approximately 40–50 per cent of the dry weight of the wood.

The fungus causes severe darkening of the wood, which breaks up into small rectangular pieces on drying. There is usually a thin skin of sound wood left on the surface of the timber but rarely is there much evidence of fungal growth.

Since this fungus requires relatively wet timber and does not possess well developed conducting strands, its eradication is much more simple than in the case of dry rot; growth will be checked at once if the timber is thoroughly dried and the source of moisture removed. Badly decayed wood, however, should be cut out and replaced with timber treated with a fungicide, and if there is any possibility of the dampness persisting, the existing sound timber left *in situ* should also be treated with a suitable preservative.

Waste disposal

The hygienic disposal of waste materials is extremely important in the control of most pests. The accumulation of food waste and greasy or sticky paper may attract rats and mice and be the breeding place for many insects. The waste should be kept in tightly covered bins or plastic sacks during its immediate accumulation at places inside the building, eg maids' service rooms. The bins or sacks should be removed by the houseporter to the main waste collection area outside the building from where they will be removed by the local authorities.

Rats and mice make nests in stores of paper, boxes, old linen and similar articles. Any accumulation of these should be moved from time to time to ensure that rodents are not making a nest.

The contents of sanibins should be disposed of in an incinerator or by chemicals in a special container.

The contents of the waste paper bins should be collected in paper or plastic sacks, while those of the ashtrays should be collected separately in bins because of the fire risk. Later both types of waste will be taken to the main waste collection area outside the building.

Rubbish chutes for dry waste and waste disposal units for food waste are found in some establishments.

Liquid waste from sinks, baths, lavatory basins, WC pans etc, is taken by a system of pipes into the house drain and so to the sewer. After emptying a sink full of dirty water sufficient clean water should be run so that the trap below the sink contains clean water.

Kitchen waste consists of various materials and in large establishments bones, fat, articles made of glass etc, are often kept separate from ordinary food waste and are sometimes sold. Food waste may also be sold for pig food.

C

ROOM INTERIORS

12

CERAMICS, GLASS, METALS, PLASTICS AND SANITARY FITMENTS

These materials are found throughout different establishments in many forms and so are considered here before the main parts of room interiors.

Ceramics

Ceramics are basically clayware. During manufacture different types and different proportions of clays, with other ingredients, are mixed to produce the ceramics of the required quality for floor and wall tiles, drain-pipes, sanitary fitments, vases, cooking utensils, crockery and the like.

The required ingredients are mixed with water to form a liquid mixture which, after various refining processes, becomes sufficiently plastic for the clay to be shaped into hollow or flat ware, either on a potter's wheel or by moulding. The clay body is then fired at a high temperature and changed into hard 'biscuit' ware. After this a thin film of glaze is applied to the surface of the article and it is then fired a second time. Decoration, when necessary, is applied underglaze or overglaze but gold, because it will not stand up to the high temperature of firing, is generally put on overglaze and thus, of course, it is more liable to be harmed and removed when the article is washed.

The word 'china' is a broad term which covers all 'clayware' used for crockery and sanitary fittings, and includes glazed and vitrified earthenware, bone china and porcelain.

Glazed earthenware

This is the ceramic most frequently manufactured in the UK. It contains a large amount of ball clay amongst its ingredients, is rather thick, opaque and the glaze is necessary as the clay body is porous.

Vitrified earthenware

Also known as vitreous china, this has extra flint added to the clay mixture and has an extra firing, when more complete fusion takes place. It results in heavier, stronger and less easily chipped, but more expensive articles.

Bone china

This contains more china clay (kaolin) and china stone than earthenware and, in addition, calcined bone; bone makes the clay easier to work and gives the body strength. Bone china is fired at a higher temperature than earthenware, when it becomes almost completely fused and non-porous; it is thin but very strong and more costly.

Porcelain

This 'hard-paste china' contains no calcined bone. It is extremely hard, translucent and expensive, and only very small quantities of it are made in the UK.

Care of crockery

Self-coloured crockery has sometimes had the clay body dipped in colour and not had the colour incorporated in it. White china with a simple, easily repeated pattern is therefore probably the most suitable for hotels and other establishments; narrow or broad bands of colour are frequently used for decoration, while badging is a deterrent against pilfering except by the souvenir hunter.

Cup and jug handles which are moulded on to the clay bodies are called sanitas handles and are used for 'hotel ware'. They are less easily broken than those stuck on but are rather clumsy in appearance. 'Hotel ware' plates have a rolled edge, to make chipping of the top of the rim less likely.

Whether washing up crockery by hand or machine the articles should be:

scraped of food scraps etc;
pre-rinsed if possible;
washed in hot water containing detergent, for hand washing up approx
   44°C and for machine washing up 60°C;
rinsed in hot water at least 77°C;
drained and air dried.

If necessary, a chemical sterilant may be added to the rinsing water.

Ceramics, metals and plastics are used in the manufacture of sanitary fitments and these are dealt with at the end of the chapter.

Glass

The main ingredient of glass is sand, which needs to be as free from impurities as possible, and to this, other chemicals are added in proportions, depending on the quality or type of glass required.

The carefully measured mixture, known as the batch, is fed into a furnace where it is heated to a very high temperature, 1300°C or more. From the furnace the molten glass is led away for shaping, after which the glass article has to be very carefully cooled. This is done by annealing: the glass travels on a conveyor belt through an annealing oven, and, after the initial reheating, the glass gradually cools as it passes through.

Glass is required for many purposes: tableware, cooking utensils, bottles, vases, lamps, doors, windows, mirrors etc, and some of these require glass with specific properties.

Soda lime glass is used for ordinary, inexpensive, flat or hollow glassware and the main ingredients are sand, soda ash and limestone.

Lead crystal glass is used for expensive hollow glassware. It is attractive to look at, has brilliance and a fine lustre. It consists of sand, red lead and potash. These ingredients produce a slightly softer glass than soda lime, enabling lead crystal to be cut more easily.

Borosilicate glass is used for ovenware since it is very hard and has special heat-resisting properties; it is the borax content which cuts down the rate of expansion when the glass is heated. Borosilicate glass could be used for flameware but in the UK flameware is manufactured from toughened or tempered glass.

Flat glass

This is used for windows, shelves etc, and is made from soda lime glass. It is of two main types: sheet and float. Polished plate glass is now largely being replaced by float glass.

Sheet glass is drawn continuously from the molten mass and passed through an annealing tower, after which it can be cut into the required lengths. The thickness of the glass can be varied as the quicker the sheet is drawn off, the thinner the glass will be; frequently there are flaws in the glass so that a certain amount of distortion occurs. However, this is happening less as production methods improve. It is used for ordinary window and picture glass.

Float glass is, in many cases, replacing plate glass as, unlike plate glass, it does not require to be ground and polished after annealing. Both are made from very refined ingredients and provide clear, undistorted vision and are used for shop windows, mirrors and protective coverings on furniture. The edges on mirrors are often bevelled, ie cut at an angle, while those on shelves are ground and the corners rounded.

Neither sheet nor float glass will allow the passage of ultra violet light, but it is possible to obtain a special window glass through which ultra violet light will pass.

The problem in cleaning flat glass, eg windows, mirrors, protective glass coverings on furniture etc, is to produce a shine without smears. After removal of dirt with damp lintfree cloth, scrim or newspaper (newsprint contains a solvent useful for windows), greasy finger marks may be removed with vinegar and water and more stubborn marks with methylated spirit. The shine is produced by a linen, lintfree cloth and 'elbow grease'.

Obscured and safety glass

Both obscured and safety glass are made from sheet or float glass.

Obscured glass required for bathrooms, and other places where light but not transparency is required, often has a pattern on one side. This is produced when the molten glass flows from the furnace between embossed rollers.

Safety glass may be:

• *obscured glass with wire*, which is incorporated during the rolling process. The wire prevents the glass falling when broken and for this reason the glass is used in doors and skylights.
• *laminated glass*, made in the form of a sandwich of two thin layers of glass with a filling of vinyl-type plastic in the middle. If the glass is broken, it adheres to the interlayer.
• *toughened glass*, made by subjecting the glass to a temperature just below softening point and then cooling the surface layers very rapidly. In this way a skin is formed on the glass and if it is broken the glass shatters into very small and comparatively harmless fragments.
• *toughened and laminated glass*, which incorporates toughening and laminating and is five times stronger than other types of safety glass.

Hollow glassware

Hollow glassware is produced by blowing, moulding and pressing; even with blown glass, cast moulds of wood or iron are often used for shaping the glass. The moulds may be patterned, giving 'imitation cutting' which is very even, and the cut edges are relatively smooth. It is sometimes possible to see mould marks on the glass left by the hinges. Pressing can only be carried out where the top of the container is wider than the bottom, to allow the plunger to be withdrawn. In this method, the molten glass is pressed into all parts of the mould by the plunger.

Cut glass

Glass is cut by hand using abrasive wheels which rotate at great speed. The cuts at first have a matt surface, but later the article is polished, often by treating with acid, and the cuts on lead crystal glass produce prismatic bodies which give beautiful colours to the glass. Cut full lead crystal glass is expensive, and used for chandeliers, decanters, vases and beautiful table glass. Engraved patterns are made with copper wheels, on to which an abrasive is fed, and these have matt surfaces and are rarely polished. Etching is a further way in which glass can be decorated; in this case the article is coated with a protective wax and the pattern is cut into the wax with a steel needle. On immersing the article in a bath of acid, the acid eats into the unprotected (patterned) areas. This method is a usual way of badging glass which may also be done by engraving, sand blasting or applying an enamel transfer which is fired.

Metals

Silver

Silver is a relatively soft metal which is found naturally in the earth, but more generally in the form of silver salts, from which the metal is extracted. It is a white metal and is unaffected by water, pure air and the majority of foodstuffs. Sterling silver is an alloy containing 92.5 per cent silver and the remainder is substantially copper, which is added to harden the silver and yet not change other properties of the metal. Sterling is obviously more expensive than silver plate and is seldom used in hotels and other establishments.

Silver-plated ware is made from blanks or bodies of a nickel silver, or nickel brass alloy. These are immersed in a complex solution of silver salts, and by means of electrolysis, silver is transferred to the blanks and an electroplated article results. This process gave rise to the symbol 'EPNS', meaning electroplated nickel silver.

Provided that the blanks are made from a specified nickel silver alloy of the correct thickness, the quality of silver plate is dependent on the electro deposit which must adhere well, be free from defects and be of a good thickness. The silver deposit may be checked out at an assay office. The deposit is relatively soft and can be scratched or abraded by repeated rubbing; in either case, in time, the base metal will eventually be exposed.

Silver-plated ware can be re-plated but the process is an expensive one, as the old deposit has to be stripped off during the preliminary treatment and if repairs are necessary or desirable at the same time these will obviously increase the cost. However, good quality silver plate is not cheap in the first instance, and may be excluded from consideration where initial cost and labour costs are prime factors; nevertheless, there is no counterpart to silver in terms of appearance and elegance.

The tarnishing of silver is due to the action of compounds of sulphur, present in industrial atmospheres and in certain foodstuffs such as eggs, onion juices, pickles, etc. The tarnish is silver sulphide and varies in colour from yellow, through brown to blue-black, depending upon its thickness. Tarnish is unaffected by simple washing operations but can be removed by the use of:

• *'silver dips'* which are based on an acid solution of a thiourea compound into which the articles are dipped (not steeped) and then washed and dried; no friction is needed. The liquid attacks stainless steel and it should only be used in a glass, earthenware or plastic container;
• the *polivit method* in which the silver articles are immersed in a hot soda solution containing a sheet of aluminium for up to 10 minutes, during which time a chemical exchange process takes place; the articles are then removed, washed and dried.

After using either of these methods the silver should be polished to restore its shine. Tarnish may be removed and the silver restored to its original shiny finish by the use of:

• *proprietary preparations* based on precipitated whiting and jeweller's rouge. The polish is rubbed on the article, allowed to dry and buffed off, followed by rinsing and drying. One of these preparations is 'long term' silver polish. This forms a very thin, colourless, transparent and relatively impervious film which is chemically bonded to the silver. The film has no odour, taste or other detectable property and in no way affects the appearance of the treated article. The film can be broken down and removed by mechanical action or abrasion. 'Long term' polishes are expensive;
• a *burnishing machine* in which highly polished steel balls and the silver articles are immersed in a detergent solution; the machine rotates and friction

is applied to the articles by the steel balls; the articles should afterwards be washed and dried. This method is particularly suitable for large quantities of silver. The articles should be carefully positioned and those with handles of ebonite or similar material are not suitable for the burnishing machine.

Steel

Steel is iron containing a little carbon and small quantities of other materials and is often used in the form of pressed steel for baths, sinks etc. To prevent corrosion it is normally coated with enamel. This is a smooth surface which scratches to varying degrees and may become stained. A liquid scourer (not a coarse abrasive) should remove any stains and, after removal, washing and drying should be sufficient to retain a clean appearance.

Chromium

Chromium is the coating on steel used for taps, bath handles, shower fitments etc. These can become water spotted and greasy but will not tarnish. Washing and rubbing up should be sufficient to keep the shine. Besides being coated with enamel or chromium, iron or steel may be:

   zinc coated, ie galvanized, for buckets etc,
   nylon or *plastic* coated, eg metal legs on furniture,
   painted, eg pipes coloured for identification purposes.

Anodizing and lacquering are other methods used to prevent tarnishing.

Stainless steel

Stainless steel is steel to which 8–25 per cent of chromium has been added, making it corrosion resistant.
   It is a tough, durable metal and usually has a mirror-polished or satin finish. It is used for sinks, WCs, hospital equipment (bedpans, bowls etc), cutlery and occasionally wall tiles. For spoons and forks, a steel containing 18 per cent chromium and 8 per cent nickel is generally used.
   Stainless steel can get scratched and it must be recognized that stainlessness is a relative property. Stainless steel can be harmed or stained by:

   silver dip solutions,
   chlorine type bleaches,
   salt/vinegar mixtures.

In addition, if wet knife blades are left in contact with galvanized or aluminium articles, eg draining boards, pans etc, staining is brought about by the deposition of a zinc or aluminium corrosion product on the steel by electrochemical action.
   Apart from avoidng the dangers mentioned above, stainless steel only needs washing and drying in the normal way.

Brass

Brass is used for door handles, occasionally stair rods, taps, orna-ments etc. Many of these items may be lacquered, when tarnishing is avoided. Unlacquered articles need polishing with a brass polish (see page 114).

Copper

Copper is more frequently found as utensils in the kitchen than in the house-keeping department. It may be found in bars in sheet form as a wall surface or counter top, when it will be lacquered, and there may be copper vases and other ornaments in the establishment. Unlacquered copper needs polishing (see page 114). There is a 'long term' polish for hard metals, ie brass and copper.

Plastics

To many people plastic conjures up a picture of a beaker, a washing-up bowl, a laminated plastic table top or a plastic bag. However, it must not be forgotten that plastics are used for adhesives, protective coatings to metal and wood, sanitary fittings, tableware, fabrics, wall and floor coverings and many other purposes.

Plastics are a group of many substances with similar, though not identical, properties. Thus, they are

light in weight,
less noisy than many materials,
resistant to most chemicals,
non-conductors of electricity,
scratchable with harsh abrasives and sharp articles,
easy to clean,
non-absorbent, but some are thermoplastic and absorb grease,
not liable to attack by moth or other pests.

Polyvinyl chloride

This has many uses. Plastic floor finishes in tile and sheet form are generally based on polyvinyl chloride and it may be incorporated with inert fillers, pigments and plasticizers to give a homogeneous mixture, or it may form a surface layer on some suitable backing (see page 171). The durability and ease of cleaning of these floors is dependent on the proportion of PVC present. It is used for wall coverings in the form of tiles, flexible sheets or as a plastic surface on some wall papers. It is also used for soil and waste pipes, electrical conduits and translucent ceilings.

Polystyrene and other foams

Polystyrene and some others, eg polyurethane, can be produced as a foam and this, when set, may be used in tile or sheet form on walls or ceilings to give heat and sound insulation, but there is a considerable fire risk.

Polyethylene and *polyurethane* are also produced as foams which have resilience and can be cut into the required sizes for mattresses (see p. 230), and into different shapes for upholstered furniture. There is a considerable fire risk with some plastic foams when toxic fumes are produced and although these materials can be treated against this hazard, the price is considerably increased.

Polyurethane is also used as a clear seal on wooden floors and furniture.

Acrylic sheet 'perspex', and reinforced plastics of the polyester/glass fibre type, are used in the manufacture of furniture, baths, showers and other sanitary fittings. They are very strong, yet light in weight (a 'perspex' bath weighs about 10 per cent of the weight of a cast-iron one); in tall modern buildings where a number of bathrooms may be built one above the other, the weight of the sanitary fittings can be of importance.

Laminates

Melamine, phenolic and other *plastic resins* are used to produce plastic laminates. These are thin veneers marketed under many trade names, eg Formica, Warerite, Duralam; they may be stuck direct to the wall, to plywood or similar supporting material and used as wall panels, counter tops and in the manufacture of furniture. The laminates are manufactured by subjecting layers of paper impregnated with plastic resins, such as phenolic and melamine, to great pressure and a high temperature and in some cases they now have textured surfaces.

Synthetic fibres

Polyamides, polyesters and *acrylics* etc, may be produced as fibres or long filaments and woven into textiles. These synthetic fibres, owing to their great strength and poor absorbency, are durable, easy to clean and quick to dry (see p. 208). They are used extensively in carpets, curtains, upholstery, bedding and uniforms. For some articles the plastics are 'bulked' to render them more fluffy and wool-like.

No doubt more plastics and further uses will be found in the future, but one particular point has become clear over the past years, which is that plastics should be considered on their own merits and not as substitutes for natural substances.

Plastics can normally be maintained just by dusting and wiping with a damp cloth or washing in hot water and synthetic detergent.

Sanitary fitments

Ceramics, metals and plastics are used in the manufacture of sanitary fitments, which are found throughout all establishments and in the housekeeping department in particular (see the table on page 159). These materials provide the smooth, non-porous, easy-to-clean surfaces required by the fitments, enabling them to carry away soil and waste water efficiently and hygienically. To prevent scratching and to retain the smooth surfaces they should be cleaned with liquid detergent or a fine abrasive (scouring liquid). Any mild acid (lemon juice or vinegar) used to remove hard water deposits should be thoroughly rinsed away or the glaze will be harmed.

Sanitary fitments

Material	Fitment	Characteristics
vitreous china	lavatory basins, sinks, sluices, WCs, urinals, bidets	can chip and craze, making surface more difficult to clean and more susceptible to staining
stainless steel (usually 18/8)	lavatory basins, sinks, draining surfaces	satin finish shows scratches less than mirror finish, damaged by chlorine bleaches, silver dip solutions and salt/vinegar mixtures
vitreous enamel coated pressed steel	baths, sinks and draining surfaces	does not retain heat as well as plastics, marks with dripping taps. May chip and rust
vitreous enamel coated cast-iron	baths	heavy, tough finish with a high gloss, if damaged rusting occurs
plastics eg acrylic (Armacast is tough acrylic sheet backed with polyurethane reinforcement)	lavatory basins, baths, sinks	light weight, retains heat and resists most stains. Cheaper ones may flex and give problems with crazing. Damaged by solvents eg paint stripper, perfumes, after-shave, and the heat of cigarettes. Can follow cleaning with fine abrasive or with metal polish

WCs and urinals

Water closets have a seat and a lid normally made of plastic; the pan may be on a pedestal clamped to the floor, or cantilevered, and in conjunction with it there is a water cistern (water waste preventer).

WC pans become easily stained and require regular and thorough brushing with a lavatory (toilet) brush. Toilet cleansers which are acidic are available and may be used when brushing is insufficient. If it becomes badly stained, the water in the trap should be removed and a fine abrasive used with friction. Concentrated hydrochloric acid may be used *under strict supervision*, but this is a job more for the maintenance department than for the housekeeping staff.

In and around *urinals* an accumulation of lime scale, urine salts and iron salts may occur and stronger acid cleansers than lemon juice or vinegar may be required for their removal.

Baths and basins

Wash basins may be on a pedestal or cantilevered, and these parts as well as the mirror, shelf and the surround, which may be of ceramic tiles, need attention during cleaning of the basin.

Baths and basins have an overflow and this, together with the plug, plughole and soapwell, must be kept free from dried soap, scum and fluff. Dripping taps should be reported for maintenance as soon as possible or a hard water stain results, and when this occurs it should be removed with a weak acid, eg vinegar or a cut lemon, and if this is not effective a stronger acid such as oxalic may be used under the supervision of the *housekeeper*. There are also proprietary substances sold for the removal of hard water marks. After the use of any acid the surface must be rinsed thoroughly or the acid will harm the glaze.

Cleaning

Chrome fittings on any fitments need washing and rubbing up to remove water marks and grease and to keep their shine.

To clean a wash basin

1 Remove hair, fluff etc, from waste, chain and overflow.
2 Wash and dry toothglass.
3 Clean basin and surrounds with swab and scouring liquid, paying particular attention to underneath and round the base of the taps.
4 Rub up taps and dry basin.

Baths and bidets

Baths are cleaned in a similar manner to wash basins, including the surround, but there is more likelihood of scum and staining, making cleaning more difficult.

If there is a shower, the fittings and curtain rail should be cleaned; the curtains should be wiped and left hanging inside the bath.

A *bidet* is made of vitreous china and has chromium-plated fittings. There is a tendency for staining by hard water and a bidet is cleaned in a similar way to a wash basin.

To clean a WC

1 Flush pan, brush well and flush again.
2 If pan is still stained, use toilet cleanser and allow time for it to work.
3 Brush and flush again.
4 Wipe pedestal, seat, lid and surrounds with a suitable cloth and dry.
5 Check for toilet paper and leave a spare.

13

FLOORINGS OR FLOOR FINISHES

Floors are important areas which are readily noticed on entering a room or particular area and they may be both functional and decorative. They cover a tremendous area and are subjected to a great deal of wear and tear. They play a very large part in the cleaning and maintenance programme of any establishment. In order that floors should remain in an hygienic condition, and retain as good an appearance for as long as is possible, some knowledge of the various types of floor finishes, their advantages, disadvantages and maintenance is necessary.

Floors frequently form the basis on which the rest of the décor is planned, outlasting other furnishings and decorations, and clean, well kept floor surfaces will often indicate the standard of cleanliness throughout the establishment.

It is essential that the floor finish chosen for any particular place should be in keeping with the purpose of the room. There may be places where durability and hygiene are of more importance than appearance, for instance in the kitchen, however, appearance is of prime importance in the lounge where an impression of warmth, comfort and quietness is expected; there is no ideal floor surface for all areas.

Choosing floorings

Carpets are regarded as soft floor coverings rather than floor finishes (see Chapter 14) but when selecting either floorings or carpets consideration should be given to their:

- **appearance**, when colour, pattern and texture play a large part,
  - pale colours, especially blues and greens, as well as shiny surfaces give a cool or cold appearance,
  - intense colours (red, orange etc) and matt surfaces give an impression of warmth,
  - the sense of scale in a room or area is influenced by the size of the pattern,
  - patterned surfaces tend to make a large room appear smaller,
  - plain colours make a small room appear larger,
  - patterns and some colours do not show spillages and soiling readily and the floorings retain their appearance of cleanliness longer than others;

- **comfort**, which is of importance to guests and staff, soft resilient surfaces are generally comfortable to walk on but may prove extremely tiring to people continually walking on them eg housekeepers,
  - the harder, noisier, colder floorings offer less heat and sound insulation,
  - noisy floorings can cause disturbance, and hence discomfort, to the occupants of a room and to those in adjacent rooms,
  - slipperiness may lead to discomfort but may often be due to the maintenance given to the flooring rather than to the flooring itself; a very shiny flooring looks slippery;
- **durability**, without due consideration to the wear and tear expected in an area a flooring may become 'tired' looking very quickly,
  - grit cuts into some floorings more easily than into others,
  - spillages of water, grease and food acids are more likely in some places and will harm certain floorings,
  - cigarette burns, dragging of furniture and the use of trolleys occur more frequently in some areas than in others,
  - areas of more concentrated wear need careful consideration, eg the foyer with well defined traffic lanes to the reception desk, lifts etc, and areas where people turn and their feet are ground into the flooring, eg bars, dressing tables, waiters' stations etc, may show excessive wear;
- **life expectancy**, a flooring needs to be durable for the length of life expected and this is not the same for all areas,
  - in kitchens, hospital wards etc, life expectancy may be for many years,
  - a bedroom floor surface, owing to changes of décor, may not be expected to last more than seven or eight years, or a bar flooring more than two or three years, especially if it is of a contemporary nature;
- **safety**, which is of great importance to all occupants of the building,
  - surfaces should have non-slip qualities when wet and dry,
  - overpolishing may cause slipperiness;
- **ease of cleaning**, which is an important factor in the running costs of any establishment,
  - the extra initial cost of a flooring which is easier to clean may be saved over a comparatively short time,
  - a flooring which is easy to clean does not necessarily maintain its clean appearance throughout the day,
  - floorings cannot usually be constantly cleaned, however easy they may be to clean, so for a well maintained flooring the flooring material, colour and pattern must be carefully selected;
- **cost**, which may limit the choice of flooring. The true cost of a flooring is the initial cost, including laying, plus estimated maintenance costs.

Subfloors

The effective life of most floorings will depend on how they were laid initially and on their subsequent care and cleaning. Many properties of a flooring may be enhanced or ruined by the base, ie the subfloor, on which the flooring is laid.

In large modern buildings the *subfloor* is made of concrete, but in older and smaller buildings it consists of soft wood boards, at least 10 cm wide nailed to wooden joists.

Softwoods are obtained from the cone-bearing trees, eg European Redwood (deal), Spruce, Douglas fir and Pitchpine. Deal is the most frequently used soft wood for these boards but it is:

poor in appearance,
not very hard wearing,
seldom left white and uncovered,
liable to warp and shrink, causing creaking and squeaking,
subject to dry rot unless adequately ventilated. Air should be introduced into the walls of the building, by mean of air bricks, to prevent the growth of the fungi which causes dry rot.

Where a deal floor is only partially covered with a carpet, the boards may be stained and varnished and left as a surround, but they are more usually completely covered with a carpet or floor finish. Even with careful seasoning or kiln drying these boards often warp and shrink, leaving gaps through which draughts and dust will rise. Central heating can also lead to shrinkage.

Concrete subfloors have the advantage of:

being solid and fire retardant,
allowing the use of underfloor heating,
not creaking,
not requiring ventilation.

There is, however, a risk of rising damp when concrete floors are placed in direct contact with the ground, and a screed of suitable material should be applied before many of the floor finishes. Ordinary concrete floors are dusty and difficult to clean, and where they have to be left uncovered they should be treated with a dust preventative, eg sodium silicate. Aerated concrete is available to increase sound and heat insulation.

A flooring should be provided with a true, level and dry subfloor. Where the finish is in tile form, the tiles should be laid evenly and close together, so preventing dirt and bacteria accumulating in the crevices. This type of flooring has the advantage that individual tiles which have become worn or damaged can be replaced. A floor finish laid *in situ* normally presents a jointless flooring and, for easier cleaning, may be continued up the wall to give a coved skirting.

General care and cleaning of floorings

Once a floor finish or flooring is laid the treatment it receives is of tremendous importance, in order to prevent the penetration of dirt and to provide an easily maintained surface. It is for these reasons that many floorings today are sealed and/or polished (see page 114).

A seal is applied to a clean, dry floor and gives a non-absorbent, semipermanent gloss or finish which will wear in time. Before the floor can be resealed, any remaining seal has normally to be stripped off; this can be done in the case of wood and cork floorings by sanding, but in other cases a chemical stripper generally has to be used. These chemicals are fairly drastic in their action, and before using a seal the reaction of the flooring to the stripper should be considered. Most seals last at least 1–2 years and the two pot plastic seals 2–4 years. In order to preserve the seal for as long as possible a polish may be applied to sealed floorings. Polishes are also applied to unsealed floorings when they prevent the penetration of dirt and spillages.

Floor finishes are, in general, harmed by either spirit or water and the choice of polish rests on this fact, as polishes are either spirit or water based. Spirit-based floor polishes may be paste or liquid and require buffing when dry to produce a shine; water-based polishes, which may be water/wax or plastic emulsions, are liquid and dry to a shiny surface which in some cases can be improved by buffing, and in others cannot. It is this ability to dry shiny that has confused water based or self-shining emulsion polishes with floor seals (see p. 117).

The amount of cleaning required by any flooring will depend largely on the amount and type of traffic it receives, but some form of daily cleaning will be necessary, while special and periodic cleaning will be required at less frequent intervals.

Daily cleaning entails removing:

1 dust and dirt by sweeping, mopping, vacuum cleaning, damp mopping or washing according to the type of flooring;
2 resistant marks, normally by rubbing with a damp cloth and a fine abrasive;
3 stains which should be removed as soon as possible, because on drying they become set and are much more difficult to remove.

Special cleaning may involve:

1 removing dirt by scrubbing or spray cleaning;
2 improving polished appearance by buffing or spray buffing;
3 re-polishing.

Periodic cleaning of polished floors involves:

1 stripping polish and possibly seal;
2 re-sealing;
3 re-polishing.

Where a floor is sufficiently large and sections are subject to heavy wear, so needing to be treated more frequently than the whole floor, *spray buffing* may be used as a fill-in system.

Dust and tracked-on soil should first be removed by sweeping, and dry or damp mopping. Diluted emulsion polish or a water-based ready-to-use compound is then sprayed lightly onto the floor area requiring it and the sprayed sections are buffed with a floor machine until dry. The adjoining areas should be dry buffed with the same pad to ensure uniformity of appearance. After spray buffing the floor is dry mopped to pick up any residue. The elimination of scuff marks, scratches and signs of wear is brought about by the re-emulsifying action of the spray and the tendency for the heat from the pad friction and the weight of the floor machine to 'melt' the surface of the emulsion polish. Spray buffing is only suitable when the flooring is protected with a good coating (2–3 thin coats) of metallized polish. It is not a method of cleaning a dirty flooring and there is a possibility of dirt being carried into the polish during the re-emulsifying action of the spray.

Therefore, *spray cleaning* is sometimes preferred. In this case a dilute solution of a neutral detergent is sprayed onto the floor and a floor machine with a green pad used to 'scrub' the surface. The pad absorbs the dirt and should be changed frequently. The floor is buffed until dry and dry mopped.

To prevent accidents and further damage to the flooring:

Loose edges should be attended to immediately;

Metal strips should be placed over the edge of the flooring at doorways, staircases, etc;

Spillages should be wiped up as soon as possible;

Excess water and polish should be avoided during cleaning. While a flooring material may be unharmed by water, the adhesive used with it may be damaged and cause the lifting of the flooring, particularly if in tile form;

In large areas 'wet floor' notices should be displayed when floors have been washed.

Classification of floor finishes

Having decided what requirements are to be met by the floor finish, it is possible to consider the types of finishes which meet the needs of a particular area.

Floor finishes may be classified according to their hardness, porosity or the material from which they are made. None is ideal and here they will be classified as hard, semi-hard or soft. A choice may therefore be made from those in the table below.

Floor finishes

Hard finishes	*Semi-hard finishes*	*Soft floor coverings*
cementitious eg terrazzo and granolithic	thermoplastic tiles	carpets – woven, tufted, adhesively bonded, electrostatically flocked (see Chapter 14)
stone eg marble, slate in slab form	vinyl in tile or sheet form	
ceramic tiles – quarries and more decorative hard-glazed tiles	rubber in tile or sheet form	
resin floorings with or without vinyl or marble chips set in	linoleum in tile or sheet form	
bitumastic laid *in situ*	cork in tile or sheet form	
magnesite and other composition finishes laid *in situ* or as small blocks		
wood – hardwoods laid as strip, block or parquet		

Hard floor finishes

Hard floor finishes are in general durable but noisy; with the exception of wood they are:

Figure 13.1   Slate staircase

  cold in appearance and feel,
  vermin proof,
  impervious to dry rot,
  fire retardant,
  easily cleaned.

Where laid *in situ* they may have coved edges to facilitate cleaning.

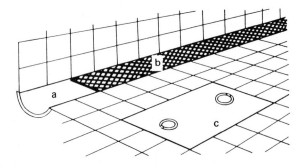

Figure 13.2   a) half-channel drainage pipe with b) metal covering, c) precast slab with lifting rings

In both granolithic and terrazzo floorings the cement is absorbent and strong alkalis should be avoided in cleaning to prevent the dissolved solids, eg soda, exerting a pressure and causing the surface of the flooring to crack or pit. It may be marked by oil and spillages, eg ink and beverages. (See table on page 176 for cleaning.)

Cementitious floorings

Granolithic	Terrazzo
• consists of granite chips set in cement • is laid in plastic state on solid subfloor • is heavy duty flooring • is used for basement corridors, store rooms, stairways and laundries • may be machine ground to produce a smooth surface more attractive in appearance and used in cloakrooms, corridors and staircases • for staircases should have abrasive material set into edge of each step to prevent slipping	• consists of marble and other decorative chips set in cement • is machine ground to produce a smooth surface • is laid *in situ* or as precast tiles, when marble pieces may be larger • when floor area is large the surface is often divided into sections with brass or ebonite strips to improve appearance • can be attractive flooring if well maintained • its variegated pattern can camouflage dust and dirt to some extent • is harmed by strong alkalis and acids and when used round urinals in male cloakrooms stains may give problems • is used for entrance halls, cloakrooms, staircases and operating theatres (is anti-static)

Stone

Marble floorings are laid in slab form and are very expensive although a cheaper form, travertine marble, is also used. This marble has small cavities in it which offer some slip resistance but allow dust and dirt to collect. Marble is obtained in white, black, green and brown colours and is used for foyers and luxury bathrooms.

Other stones used as floorings are sandstone, quartzite and slate. They are all hard wearing and obtainable in a variety of colours. Sawn or polished finishes tend to be slippery but riven finishes are non-slip.

Ceramic tiles

Ceramics tiles are clayware available in a great variety of qualities, colours and sizes.

Quarry tiles are made from a natural type of clay, often of several blends, and are fired under pressure to make them hard and durable. Different qualities are produced, and the harder tiles are less absorbent but more slippery; however, it is possible to obtain tiles with slightly abrasive surfaces so rendering them less slippery. The tiles may be of various thicknesses and are generally 10 cm, 15 cm or 23 cm square, and red, yellow, buff or blue in colour. They should be laid close together to prevent dirt and bacteria accumulating in the crevices. Coved tiles and removable pre-cast slabs (to cover service pipes) are available to facilitate cleaning and maintenance. When laid properly quarry tiles form an impervious, hard-wearing surface and are used in cloakrooms, kitchens, canteens and any place used for the preparation or storage of food.

Ceramic tiles with a particularly hard glaze and a wide range of colours are used as more decorative floorings. They are often used in bathrooms of the more luxurious type, patios and similar places in colours harmonizing with the wall tiles. Tessellated tiles are small ceramic tiles often used as mosaics, giving a highly decorative floor.

Ceramic tiles are not affected by water, grease, acids or alkalis but the grouting may be, and so strong alkaline cleansers should be avoided. They may crack or break with heavy weights and marks can be difficult to remove if left. The initial cost may be high but this must be offset by the need for little upkeep (see table on p. 176 for cleaning).

Resin floorings

These consist of synthetic resins, usually epoxy, polyester or polyurethane, with appropriate hardeners or curing agents. Vinyl or marble chippings may be included to give a more decorative flooring, resembling terrazzo. Resin floorings may be laid *in situ* (seamless) or as precast tiles.

Polyurethane floorings are the most usual and are extremely hard wearing, unaffected by spillages of water, food, alcohol and most chemicals. Due to the resilience of the resin the floorings show a good recovery to point load and in spite of the shiny surface they are non-skid.

Polyurethane floor finishes may be used in kitchens, canteens and other areas where food is handled, bathrooms, cloakrooms, corridors and laundries (see table on p. 177 for cleaning).

Bitumastic flooring

This is a jointless flooring and consists of a type of asphalt rolled on to a solid subfloor in a hot plastic state when a sluice hole may be incorporated, and it can be continued up the wall as a coved skirting. It is soft in texture although its appearance is hard and it is completely impermeable to water. It is normally black, red or brown, but may have other colours rolled in, giving a mottled effect. It may be used in public bathrooms, hospital corridors and other heavy traffic areas. It is also used as a damp-proof membrane to protect other floorings from rising damp.

Bitumastic flooring softens with heat and dents with heavy weights, and is harmed by spirit, oil and acids but the initial cost is low (see table on p. 177 for cleaning).

Magnesite flooring

This consists of wood flour and other fillers mixed with burnt magnesite, and is laid *in situ* or in the form of small blocks. This finish is extremely porous and washing should be avoided when possible. It is therefore used where there is little risk of water being spilt, eg linen rooms. It may be sealed and/or polished to prevent the penetration of water and dirt. Magnesite flooring is moderately warm in appearance and the initial cost is low but it is limited in colour. It is harmed by water, most chemicals and coarse abrasives (see table on p. 177 for cleaning).

Types of wood floorings

Strip wood flooring	Wood block flooring	Parquet flooring
• consists of lengths of hardwood strips less than 10 cm wide • the strips are fixed to joists or to timber insets in concrete • has resilience and is very suitable for ballroom floors, gymnasia etc • extra resilience is given to a 'sprung' floor by putting springs under the joists	• consists of either (a) hard or soft wood blocks (23 × 2.6 cm or 30 × 5 cm and 2.5–5 cm thick) often laid in a herringbone pattern, or (b) panels about 45 cm square, consisting of small blocks of decorative hard woods often laid in basket pattern or as a mosaic • is laid in an adhesive on a level concrete base • is used in entrance halls, boardrooms, libraries, offices etc	• consists of rectangular pieces of wood (23 × 7.6 cm or 30 × 5 cm and less than 10 mm thick) • specially selected and kiln-dried hardwoods are used – oak, walnut, teak etc • the pieces are pinned and glued to a wooden subfloor in a herring bone pattern • is used for prestige areas, foyers, lounges, boardrooms • a cheaper but less hard wearing parquet flooring may have only a veneer of good quality wood on the surface

Wood Floor finishes

Good quality wood finishes (see table above) are among the most beautiful floorings, providing the variety of the wood and the size of the unit are chosen for effect. These floorings, which are to be mainly uncovered and subjected to a good deal of wear, must be of hard woods which come from broad-leafed trees such as oak, teak, maple, walnut, birch, beech, etc. There are varying degrees of hardness and on the whole they give better resistance to abrasive wear and indentations than soft woods. (Pitchpine, a soft wood, is however harder than some hardwoods, eg agba, African walnut.) The choice of the individual species will vary according to its colour, the scale and definition of its grain, and its rate of wear. Zimbabwean teak, muhimbi, East African olive, rock maple, missanda, birch and jarrah are exceptionally hard wearing woods. The first three are the more decorative. Birch, maple and jarrah are very suitable for ballroom floors.

The right choice of timber is very important as apart from grooves and splintering caused by bad wear, the opening up of joints may occur from wrong selection. To ensure close joints, attention must be paid to the moisture content of the timber at the time of laying. As wood is absorbent, the required minimum moisture content will depend on the degree of heating in the building, eg with intermittent heating 12–16 per cent, with central heating 10–14 per cent.

To prevent absorption of spills and dirt, wood floorings should be sealed and/or polished. The most usual seal on wood is oleo-resinous but a two-pot polyurethane is more durable. Before polishing the flooring should be free

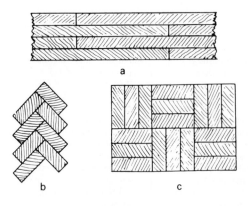

Figure 13.3 a) Strip wood flooring, b) wood block or parquet herring bone pattern, c) wood block or parquet basket pattern

from dust and dirt and any build-up of polish should be avoided. Water is harmful to wood floorings, especially if they are unsealed, and strong alkalis cause wood to disintegrate, discolour and splinter. Wood floorings can be renovated with comparative ease by sanding, after excessive wear or neglect, so that a new surface is exposed.

Wood floor finishes have a good appearance if well maintained and are poor conductors of heat, so are good insulators. Wood is resilient and wood floorings are therefore less tiring to walk on than unyielding substances. They are, however, inflammable and susceptible to dry rot; they become scratched and will splinter with the dragging of heavy articles. The initial cost is comparatively high, but if properly cared for the wood will mature and improve with age long after other materials have lost their appearance (see table on p. 176 for cleaning).

Temporary wooden floorings, consisting of movable panels, may be laid on carpeted areas for dancing.

Semi-hard finishes

Sometimes called smooth floorings, semi-hard finishes are in general:

> smooth,
> durable but normally less permanent than hard floor finishes,
> resilient (except thermoplastic tiles)
> of quite good appearance,
> unaffected by insect pests and fungi,
> relatively easy to clean.

Most are obtainable in tile and sheet form and there is a wide range of price, colour and porosity. Water-based polishes may be used on the non-porous ones and the porous ones may be sealed, then water-based polishes may be used to protect the sealed surface.

Thermoplastic floor tiles

These are made from a variety of asphaltic binders with inert fillers and pigments. They are rigid tiles, usually 23 cm square, and are laid on a clean, smooth, rigid surface, set as closely together as possible in an approved adhesive. They are laid in a warm, pliable state (thermoplastic) but harden on cooling and may be carried up the wall to form a small coved skirting. They may be polished with a water-based polish, but polishing is liable to make them slippery.

They tend to be cold in appearance and because they have no resilience they are hard and noisy, dent with heavy weights and scratch easily, ie with grit, sharp edges, coarse abrasives. As they are thermoplastic they soften with heat and they will then dent more easily.

Thermoplastic tiles are non-porous, but strong alkalis will remove their surface rendering them porous; they are also harmed by grease and spirit. Although cleaning is relatively easy they show marks badly, eg heel marks.

They are durable, obtainable in a variety of colours and are a comparatively cheap flooring; the initial cost is an important factor in their choice. They may be used in bathrooms, corridors, offices, etc (see table on p. 177 for cleaning).

Vinylized thermoplastic tiles are also available. The introduction of the vinyl resins greatly improves the colour and wearing qualities of the tile without increasing the price very much.

Vinyl floor finishes

Vinyl floor finishes are manufactured from PVC and similar synthetic resins, inert fillers and pigments. They are very much more resistant to damage than other semi-hard floorings but heat, even of discarded cigarette ends, scars the surface.

There are two main types:

1 *Vinyl asbestos* which is obtainable only in tile form. The fillers include short fibred asbestos. The tiles are rigid and similar in appearance and feel to thermoplastic tiles and are stuck to the subfloor with a suitable adhesive.
2 *Flexible vinyl flooring* which is obtainable in tile or sheet form. During manufacture plasticizers are added to render the product flexible and the sheets can be welded to form a completely impervious floor. Sometimes the basic mixture is mounted on canvas or some other suitable backing material, eg felt, foam rubber or cork, and these backing materials add to the resilience and softness of the flooring and contribute to the reduction of noise; but even without these backing materials it is more comfortable underfoot than either of the two previously mentioned tile floorings. This type of flooring may or may not be stuck down.

The greater the amount of vinyl in the flexible flooring the greater the resistance to wear, grease, scratching and indentation to point load. The vinyl content may be a surface layer only and a wear surface of 0.5 mm is considered suitable for contract work. The homogeneous sheet or tile floorings are more expensive but not necessarily superior to those with a surface layer of vinyl and PVC chips may be embedded.

Cushioned vinyl floorings are flexible vinyl floorings which contain a layer of foamed vinyl. These prove effective sound insulators and are more resilient than the other vinyl floorings. It is possible to produce textured effects when wood block, ceramic tile and rush matting and even carpet may be simulated.

Slip-resistant vinyl floorings (Altro safety floorings) are glass fibre backed abrasive sheet vinyl floorings. The main constituent is ultra high grade PVC, impregnated throughout the thickness with aluminium oxide granules and with carborundum granules in the upper section. A bacteriostat is added and gives antibacterial activity on the exposed surface.

Vinyl floorings can be used in a great variety of places, including bathrooms, corridors, canteens, offices, study bedrooms, hospital wards and corridors (see table on p. 177 for cleaning).

Rubber floorings

These are obtainable in tile and sheet form. During manufacture, rubber with filling materials and pigments is vulcanized (heated out of contact with air) to give a hard finish and the resulting tiles or sheets may be of many colours. They should be laid in a suitable adhesive on a smooth subfloor, and may be left unprotected or polished with a water-based polish.

Rubber flooring is soft, quiet, resilient and comfortable to walk on. It is extremely durable if properly maintained and a life of twenty to thirty years is not uncommon. It is non-absorbent and resists water. It is harmed by spirit, grease, sunlight, alkalis and coarse abrasives and it marks badly, especially with rubber heels. Polishing with a water-based emulsion affords some protection against sunlight and the scuffing of rubber heels. Rubber flooring may be used in bars, entrance halls, canteens, and any place where noise should be kept to a minimum.

Oil and grease resistant rubber tiles are now available (made from nitrile rubber) but the colour range is more limited.

Rubber can be used for an infinite variety of mats and mattings found in places where protection is required for the floor beneath (see page 190). Front door mats, mats in front of service lifts, and nosings on stairs may therefore be made of rubber (see table on p. 177 for cleaning).

Linoleum

This consists of a mixture of powdered cork, resin, linseed oil and pigments, put on a foundation of jute canvas and subjected to heat and pressure. The product is passed through polishing rollers and further treatment is given to harden it. It is possible to obtain factory-sealed linoleum which helps overcome the necessity for preparing and sealing ordinary linoleum.

The thickness of the mixture laid on the backing varies with the quality from 1.2–6.7 mm; 3.2–4.5 mm is most popular for most contract usage. In good quality linoleum the colour and pattern are inlaid, ie right through to the backing, whereas in cheap qualities they may only be printed on the surface, and so wear off. Linoleum may be bought in rolls, usually 182 cm wide, or in tiles. The tiles are always stuck down while the rolls may be stuck or laid loosely. In sheet form linoleum is liable to shrink, and unless stuck down the edges should overlap when laid and be trimmed later.

Linoleum is reasonably priced and extremely hard wearing (there are places where it has been down for thirty to forty years). It is subject to denting and scratching, but there is a special toughened form available which does not dent or scratch so easily. It is absorbent unless sealed and only minimum amounts of water should be used when cleaning. Linoleum is harmed by coarse abrasives and alkalis, and becomes slippery if over-polished. It is marked by cigarette ends and rubber heels, but the marks may be removed by light rubbing with fine steel wool, fine scouring powder or a paraffined rag. Linoleum may be sealed and/or polished and is used in many places, for example, linen rooms, study bedrooms, offices, corridors, bathrooms, canteens and hospital wards (see table on p. 177 for cleaning).

Cork tiles

These are made from granulated cork, moulded into blocks which are subjected to pressure and high temperature. During this process the natural resins bind the granules, and the blocks are then cut into tiles of the required size and the required thickness, usually 0.5–1 cm. Variations in the brown colour of the tiles result from the different amounts of pressure and heat to which the blocks are subjected. In addition to these natural cork tiles, which owing to their absorbent nature require sealing, there are factory waxed cork tiles, resin reinforced waxed cork tiles and vinyl cork tiles. The last mentioned have the resilience of cork and the durability of vinyl and they are more expensive.

Cork floorings have a warm and restful appearance; they are quiet and can be sanded down to expose a new surface. They are absorbent, burnt by cigarette ends and have little resistance to indentation when granules may become loosened and lost.

Cork tiles can be used in offices, corridors, bathrooms when vinyl surfaced and as surrounds to carpets (see table on p. 177 for cleaning).

Conductive or anti-static floor surfaces

These are required in computer rooms and hospital operating theatres so that there is no build-up of static electricity and floorings used may be terrazzo, magnesite, linoleum, rubber or flexible PVC. Linoleum, PVC and rubber floorings should have an electro-conductive adhesive when stuck down. To increase the anti-static properties carbon black or copper salts may be added during manufacture or wire mesh may be laid over the subfloor to improve conductivity.

Floor polish is not used unless specially agreed and then a metallized emulsion polish may be used. There must be no build-up of polish and regular tests should be carried out to ensure the level of conductivity remains within specified limits (see chart on p. 177).

It has been stated that when choosing a floor finish for any given situation, appearance, comfort, durability, ease of cleaning and the cost of the material (including laying) have to be taken into consideration.

It is not easy to compare the appearance of different finishes in the same way that comfort and durability can be compared because there are variables which may affect the appearance of the same basic material. The colour, pattern and texture all affect the appearance of the same flooring as well as its

Properties of different floor finishes

Flooring		Warmth to touch	Quietness (impact noise)	Resistance to			
				Slip	Wear	Water	Indentation
Hard							
Bitumastic		F	F	G-F	VG	VG	F-P
Ceramic tiles { Quarries, Hard glazed		VP	VP	G-F	VG	VG	VG
Granolithic Terrazzo Marble		VP	VP	G-VP	VG-F	G	VG
Magnesite		F	F	G-F	G	VP	VG-G
Polyurethane resin		F	F	G	G-F	VG	VG
Hard wood: { oak strip, oak block, oak parquet		G	P	G-F	VG-F	P	G-F
Semi-hard							
Cork tiles		VG	VG	VG	G-F	VP	VP
Linoleum		G-F	G-F	G-F	G	F	P
Rubber		G-F	VG-G	G	VG-G	G	G-F
Thermoplastic		F-P	P	G-F	F	G	P
Vinyl asbestos		F-P	P	G-F	G	G	F
Vinyl flexible		G-F	G	G-F	VG-F	VG-F	G-F

VG = very good   F = fair   VP = very poor
G = good   P = poor

component shape, eg the size and shape of ceramic tiles.

The table (left) shows some of the properties of the different types of floor finishes which can be more easily compared.

The following is an example of a cleaning specification which may be found in a housekeeping department. The example given is for linoleum but it could be amended for cleaning any of the finishes in the table on pages 176–7.

Example of a cleaning specification

Linoleum	Initial finish – sealed and water emulsion polish.
Daily	Mop sweep, damp mop or vacuum clean.
	Buff lightly to bring up shine.
Special	Apply solution of neutral detergent to floor with mop.
	Leave a few minutes (a scrubbing machine with mild pads or steel wool used by hand for small areas may be necessary.)
	Use wet pick-up or mop to remove water and soiling.
	Rinse with little water.
	Pick up.
	Allow to dry and apply polish.
Periodic	Apply stripper to floor with mop.
	Leave a few minutes (fine steel wool or mild pads under machine may be necessary.)
	Use wet pick-up or mop to remove loosened wax and soiling.
	Rinse with little water.
	Pick up.
	Allow to dry and apply 2 or 3 coats of water emulsion polish.
Avoid	Ammonia, alkaline or acidic cleansers. Coarse abrasives. Excess water.

Cleaning of floorings

Type of flooring	Daily clean	Special clean	Periodic clean	Special remarks
Stone and clay quarry tiles ceramic tiles granolithic terrazzo marble	sweep, wet mop or use electric scrubber with hot water and synthetic detergent. Wet pick up or mop and wringer bucket	frequency depends on soiling – approx. weekly. Use electric scrubber and wet pick-up	as special clean	remove stubborn marks with steel wool or fine scouring powder *terrazzo and marble* – avoid use of acids and strong alkalis. May be sealed with water-based seal
Wood polished with spirit-based polish	sweep, mop or vacuum clean. Buff with electric polisher 2–3 times a week	apply spirit-based polish and buff	(1) remove dirty wax by applying polish and 'scrubbing' with a green pad under electric machine. Change pad frequently. Buff until dry, or (2) remove polish with solvent stripper, pick up loosened wax and dirt. Re-apply spirit-based polish. Buff, or (3) in extreme cases the wood may be sanded down and the polish re-applied	
sealed wood, treated as plastic floorings				

Type of flooring	Daily clean	Special clean	Periodic clean	Special remarks
Plastic vinyl, PVC, thermoplastic sealed wood sealed cork (sealed by manufacturer)	sweep, mop or damp mop. Remove stains, eg heel marks, with neutral detergent and green scouring pad using gentle friction	(1) light scrub in heavily soiled areas by hand or scrubbing machine followed by mop with wringer bucket or wet pick- up (2) to repair emulsion polish, spray buff using diluted water-based polish on worn areas and buff with floor machine until dry. Dry mop	(1) spray clean using dilute solution of neutral detergent and a green pad under floor machine. Pad absorbs dirt and needs changing frequently. Buff until dry. Dry mop, or (2) strip polish using floor machine and alkaline detergent for non- metallized polish or highly alkaline solution especially formulated for metallized polish. Rinse thoroughly	After stripping rinse floor thoroughly with cold water and for metallized polish neutralize with vinegar in final rinse

Bitumastic as for plastic floorings but avoid use of spirit
Linoleum as for plastic floorings but avoid excess water
Magnesite as for plastic floorings but avoid water
Rubber as for plastic floorings but avoid spirit and strong alkalis. Alkaline detergent used for stripping should be removed as quickly as
    possible and the flooring immediately neutralized with vinegar and water
When terrazzo, magnesite, linoleum, rubber or flexible PVC flooring is required as an *anti-static flooring*
    soft water is recommended for mopping or scrubbing and for thoroughly rinsing
    only synthetic detergents must be used, *no soap*
    only water-based polishes (generally metallized), recommended for anti-static floorings should be used and then only if their use has been agreed
    there must be no build-up of polish
    avoid possibility of polish (particularly spirit-based) being carried on soles of shoes onto the flooring from different areas

14

CARPETS

Carpets are used extensively in all types of establishments because of their appearance, the safety factor, warmth and sound insulation. A good carpet should keep its colour, not flatten unduly with heavy furniture and withstand the expected wear and tear of traffic and spillages.

Carpets originated in the East and were all hand made; now they are mainly machine made and in some cases may be reproductions of the original Eastern ones. There are many carpets from which to choose, with a wide price range, and carpet performance may be judged by:

durability, dependent on – resistance to wear,
– resistance to abrasion,
– good construction with tufts well held,
appearance, dependent on – resistance to flattening,
– soil and stain resistance,
– colour fastness to sunlight, water, shampoo, and rubbing
safety, dependent on flame retardance,
comfort, dependent on – no build-up of static electricity,
– length and softness of pile.

Fibres

In general carpets consist of a backing or foundation and a surface pile which may be cut or uncut. A soundly constructed, firm backing is essential and this is normally made from jute or cotton threads although others, eg linen, hemp, are also used. The pile yarn may be from wool, cotton, rayon, nylon, polyester, acrilan, courtelle and polypropylene.

Wool has been the main fibre for many years because it:

has resilience,
withstands abrasion,
feels warm,
does not soil or ignite readily,
retains its appearance well if properly maintained.

Cotton – wears well,
– has little resilience,
– fades badly and loses its appearance,

– is not normally used for contract carpets.

Rayon – does not compare with wool or the synthetics,
– is a cheaper fibre,
– when introduced into a blend reduces the cost of the carpet.

Evlan, a modified rayon, has improved properties and is now more frequently used than rayon.

All *synthetic fibres*:

are resistant to abrasion,
have low moisture absorbency and stains may be easily wiped off,
melt with the heat of cigarette ends,
are electrostatic in dry conditions,
are moth resistant.

Nylon stands up to salt and sand brought in on shoes from icy roads and beaches but does not compare with wool with regard to resilience and soil resistance. A 100% nylon carpet, therefore, does not retain its appearance as well as a 100% wool carpet.

The acrylics, acrilan and courtelle, resemble wool more than nylon in handle and appearance. Acrilan dyes more easily than other synthetic fibres and courtelle takes a high twist which makes the carpets more resistant to shading and tracking.

Polyester has poor recovery from crushing but with its soft handle is very suitable for longer piled rugs.

Polypropylene has good abrasion resistance and is used in some of the adhesively bonded felted carpets.

Because of the varying properties of the different fibres, blends of fibres have become popular for carpets. These include:

wool/nylon, 80/20
acrylic/rayon, 50/50
wool/evlan/nylon, 45/40/15
courtelle/evlan/wool, 40/40/20
polypropylene/acrylic, 50/50

Carpet fibres with decreasing resistance to abrasion:

nylon, sisal, polypropylene and polyester, cotton, acrylics and wool, modified rayon, rayon.

Carpet fibres with decreasing resilience:

wool, wool/nylon, nylon and acrylics, polyester, polypropylene, modified rayon, cotton, rayon.

The type of pile is extremely important as most of the wear falls on it, but without a good backing a hardwearing pile is insufficient for a good carpet. The quality of a carpet is dependent on the:

type of fibre or fibres in it,
pile density and weight,
pile height and quality,
firm anchorage of the pile.

Carpet manufacture

This falls into two main categories:

woven   – Wilton
              – Brussels
              – Axminster
              – Oriental
non-woven – tufted
              – pile bonded
              – needle punched
              – electrostatically flocked

In woven carpets the backing and surface pile are produced together during the weaving process, but in non-woven carpets the surface pile is attached to a pre-made backing.

Woven carpets now only represent 10% of UK sales, and of the 90% non-wovens tufted carpets account for 75%.

Woven carpets

Wilton carpets

These can be made in both plain and patterned varieties. The pattern is controlled by a special device (a Jacquard) on the loom and this enables one coloured thread at a time to be drawn up as pile, while the remaining threads, 'deads', are hidden in the backing of the carpet. It is unusual for there to be more than five colours because more would mean a great deal of wastage of pile yarns carried along the backing. Two or three colours or tones of one colour are most usual.

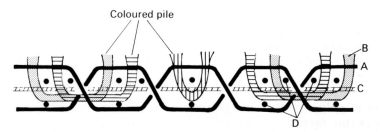

Figure 14.1   Coloured Wilton showing pile threads running along the backing where not required on the top. A chain, B pile, C stuffer warp, D double weft

Wilton carpets have a firm, smooth back with streaks of colour in it where the particular threads are not required as surface piles. The pile is cut and close, confining spillages to the surface, and is often made up of 80/20 wool/nylon. It may vary in length, even in the same carpet, which results in a textured appearance. It is sometimes made of curled yarns, giving a short twisted pile, when shading and soiling are less likely to show.

Plain Wilton carpets are made on a similar loom to the patterned, but without the Jacquard, and extra jute threads, known as stuffers, fill the back of the carpet instead of the hidden threads, 'deads'.

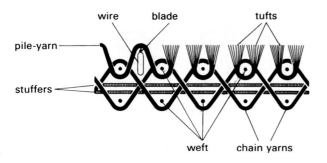

Figure 14.2 Plain Wilton

Brussels carpets

Brussels and cord carpets are variations of the Wilton weave in which the pile is uncut, ie looped. Brussels is an uncut patterned Wilton and cord is an uncut plain Wilton. The latter was originally hair cord, ie made from a mixture of hair fibres from horses, goats or cows, but it now frequently has rayon or cotton added, which makes it not nearly so hardwearing but much cheaper.

Although Brussels carpets are less popular than the cut pile Wilton they can be very hardwearing for contract use. Looped or uncut pile carpets probably give 5–10% more wear than the same quality cut pile, but they have not the same softness or resilience as the cut pile carpets.

Axminster carpets

In general, Axminster carpets are woven in such a way that the pile is almost entirely on the surface and the backing has a distinctive rib; no dead threads are carried in the backing and price for price the pile is longer and less close than in Wilton.

Spool Axminster is the most popular type of Axminster carpet and can have an unlimited number of colours in the design. A characteristic of this type of carpet is that the pattern can be seen on the reverse side.

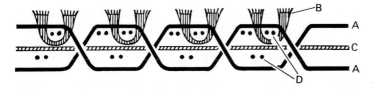

Figure 14.3 Spool Axminster: A chain, B pile, C stuffer warp, D double weft

Gripper Axminster is similar in appearance to many spool Axminsters, but owing to its method of construction the number of colours is usually limited to eight.

Chenille Axminster is quite unlike any carpet mentioned so far, but is still a woven carpet. In this case the pile is produced first as a long strip rather like a furry caterpillar (*chenille* is French for caterpillar), and during the weaving of

Figure 14.4   Gripper Axminster: A chain, B pile, C stuffer warp, D double weft

the actual carpet, catcher threads attach these strips of pile to the backing. The catcher threads holding the pile can be seen quite distinctly in the finished carpet. The result is a soft, thick carpet with unlimited colours and design. However, chenille is not produced in any quantity these days.

Oriental carpets

These are hand-woven carpets from the Middle and Far East. Genuine Oriental carpets are extremely hardwearing and the price of some will increase with age, provided they are in good condition. There are carpets and rugs from Persia, India, China, etc which are antiques and fetch very high prices.

The pile may be of wool, silk or a mixture of these and is made by the individual worker knotting lengths of yarn to the cotton warp threads of the hand loom. The type of knot varies slightly from place to place; there are, for example, Persian, Turkish and Gheordez knots (see Fig. 14.5 below).

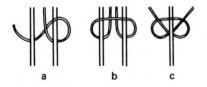

a   b   c

Figure 14.5   Typical hand-tied knots: a Persian, b Turkish, c Gheordez

Cotton threads form the weft, as well as the warp, in the backing and both vary in thickness from one carpet or rug to another. Oriental carpets are only made as carpet squares (over 213 cm × 121 cm) or rugs (under 213 cm × 121 cm), often having two fringed ends, and never as strip or body carpet.

Persian carpets or rugs were made originally as tapestry wall coverings, floor coverings on which people sat, or as prayer mats when there was a one way pattern. The patterns are generally all over, delicate, intricate and often of symbolic motifs, flowers and pine cones. Whole families, even several generations, may contribute to the making of a Persian carpet with the result that irregularities in design and dye often occur. Vegetable dyes are used, which are blended to produce a variety of beautiful subtle colours.

Indian carpets were generally made from coarser, longer pile than other Oriental carpets and less patterned than the Persian ones. There are, of course, many modern Indian ones on the market and some of these are of poor quality, tend to flatten and moult, but when they are of good quality they wear extremely well.

Chinese carpets have a close, silky pile with a well-defined pattern (not usually all over), often hand carved from the pile.

While Oriental carpets are not in normal use in establishments because of their cost, some may be found in the foyer or lounge of a luxury hotel or club, or in the manager's office. Repairs to these carpets can only be carried out by experts at great expense as the materials, colours and patterns all require skilled attention.

Machine-made copies of the genuine carpets are produced nowadays but these are not so hardwearing and will naturally cost less. They may be in use in some hotels.

Non-woven carpets

Tufted carpets

These are produced by a much faster and cheaper process than weaving. When first introduced the pile was of viscose rayon but, owing to its tendency to soil and flatten, this did not prove satisfactory and now 100% synthetic fibres and blends with wool are used, giving a wide price range.

The pile yarn is inserted into a pre-woven backing by a long row of needles and a loop is formed. This may be left uncut or cut. Mixtures of high and low piles, cut and looped areas, may be found in the same carpet. The pile is firmly held to the backing material by an application of natural or synthetic rubber adhesive; on top of this a secondary backing of hessian is added to give the carpet body and to prevent stretching and buckling, which were problems when tufted carpets were first introduced. (See Figs. 14.6 and 14.7 below.)

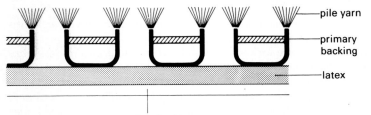

Figure 14.6   Cut pile tufted

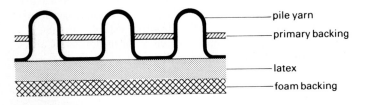

Figure 14.7   Loop pile tufted

Tufted carpets do not fray and may be cut to any shape. They may be patterned or plain and with the development of new machines there is now a wider range of patterns than formerly.

Shag pile carpets are normally coloured tufted carpets with a long, tumbled appearance pile but there is one type made in a Wilton weave.

Pile bonded carpets

These generally have a pile of nylon or polypropylene which is stuck into a PVC backing. There is a variety of techniques and either a corded or cut pile surface may be produced; pile bonded carpets are also available as tiles.

The dense pile has a firm anchorage and the carpets do not fray; seams can be bonded and the carpet may be stuck to the floor.

Needle-punched carpets

These are resin impregnated or heat-treated materials used in sheet or tile form.

A thick web of nylon, polypropylene, acrylic or polyester fibres used as 100%, or as blends, is fed into the needling sector of the machine. By the reciprocating movement of several rows of barbed needles which penetrate the web, the fibres are entangled and consolidated.

Often a supporting fabric (a scrim of polypropylene or jute) is fed along with the web into the needling machine and the fibres are pushed through the fabric, giving dimensional stability. The needled product is then either heat treated or impregnated with a resin, compressed and finally dried.

Coarse denier fibres are used for needle-punched products used for sports facilities.

Electrostatically flocked carpets

These are produced by projecting electrically charged fibres downwards into an adhesive-coated backing material.

Straight nylon fibres are locked by molecular bonding into a glass fibre reinforced vinyl backing. The 'Nylon 66' fibres are round, smooth, non-hollow and closely packed (contract range has 50,000 fibres per square inch), making the carpet (eg Flotex) hardwearing, easy to clean and quick to dry. The vinyl base is completely waterproof and the carpet is anti-static under normal conditions.

Flotex tiles have an extra heavy-duty backing which ensures that they stay flat and are suitable for loose laying. They are anti-static.

Electrostatically flocked carpets have many uses in the National Health Service and can be used in 'wet' areas, eg geriatric units, waiting rooms, canteens, entrances, cloakrooms etc, as well as in other establishments.

Fibre-bonded and *flocked carpets* may receive a Fibre-bonded Carpet Manufacturers' Association (FBCMA) classification.

Size of carpets

Carpets and carpeting are made in varying widths depending on the weave and type of carpet. *Broadloom carpets* are woven on extra wide looms, and are normally between 2 m and 5 m wide but it is possible to get one variety as much as 10 m wide. *Body* or *strip carpet* is usually 68–90 cm wide, and has no border so that the pattern matches when the strips are joined for fitted carpets (also called close fitted or wall-to-wall-carpeting), ie where the floor is completely covered with carpet. The pile of a carpet lies slightly in one direction, so care must be taken when joining the carpet that the pile all lies in the same direction. If not, it is possible to get slight variations in shade.

A *carpet square* is a loose carpet with all edges neatened; it is not necessarily a square, but should strictly be over 210 cm × 120 cm. It does not fit the floor exactly and can be turned round to even the wear. A rug or mat is normally oblong, less than 210 cm × 120 cm, has all edges neatened and is often used in front of a fireplace or beside a bed.

Stair carpet often has a border and may be 45 cm, 56 cm, 68 cm or 90 cm wide, while broadloom carpeting can be used on a very wide staircase. These variations enable the same design to be used along corridors of different widths from the stairs.

Choice of carpets

Carpets may be plain or patterned. A *plain carpet* makes a room look larger, and enables patterned materials to be used for other furnishings to better effect, but it shows dirt, stains and crush marks (ie shading due to the bending of the pile) much more readily than a patterned one. When *patterned*, the size of the design should not make the room appear small and an all-over design is probably the most serviceable as dirt, stains and crush marks will not show so easily (see p. 262).

Carpets can be woven to any particular design or colour, provided a sufficient length is being bought; manufacturers will provide dyed tufts of fibres from which a colour scheme may be selected and they will weave a trial sample if desired. Some carpet colours fade more readily than others, but with improved dyes fading is now not such a problem.

Figure 14.8   Example of the use of plain and patterned carpet

There is a tremendous price range in all types of carpets and the choice of the grade of carpet will be determined by the amount of wear and tear in a given area coupled with the life expectancy of the carpet (see p. 162). The function of the area will therefore not only determine the desirability of using a certain texture or design but also the weight or grade of the carpet. Unless it is a place where a frequent change of décor is expected, the dearer carpets are normally chosen.

A good carpet should have a firm backing and pile anchorage as well as a resilient pile.

A standard Wilton carpet has 256 warp yarns in a width of 68 cm,

an Axminster has 189,

a tufted carpet for heavy wear has 64 tufts in 12.7 cm,

a densely woven luxury carpet should have 13 rows per lengthwise 2.54 cm.

The pile height of a Wilton carpet is normally 0.5–0.64 cm,

an Axminster, 0.5–0.78 cm,

a tufted, 0.64–0.76 cm.

The lower the pile height, the closer the weave should be to prevent the backing from showing and the easier the carpet is to maintain but it is less glamorous looking.

Since 1982 pile carpets have been classified by the British Carpet Performance Rating Scheme (BCPRS). The BCPRS grades are as follows:

A – extra heavy wear   (foyers, lounges, restaurants and other public areas, corridors etc)
B – very heavy wear   (as above)
C – heavy wear   (as above and bedrooms)
D – general wear   (bedrooms)
E – medium wear
F – light wear

Before buying it is wise to send carpet samples to a testing house to have tests carried out for abrasion, dynamic loading, compression recovery and to ascertain the analysis of the pile construction (see p. 94).

Carpet laying

To obtain maximum wear from a carpet it must be well laid, and this is a job for the expert. The subfloor should be smooth and dry, with no wide cracks in it or any protruding nails.

A carpet may be stuck to the floor when it has been suggested that the carpet may last longer, but it means the carpet cannot be taken up and the good parts used elsewhere. When it is not stuck down a carpet underlay made of felt, rubber or synthetic foam is essential (see Fig. 14.9) in order to:

eliminate any slight unevenness in the floor,
retard crushing and creeping,
provide an extra layer of heat and sound insulation,
make the carpet feel soft and luxurious,
take the strain of feet and lengthen the life of the carpet.

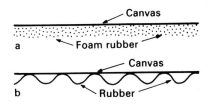

Figure 14.9 Carpet underlays

Rubber or synthetic foam underlays may be fixed to some non-woven carpets. Laying of the carpet is then easier and quicker, and installation costs are reduced, but the underlay may be damaged by damp and abrasion. Because of its great resilience it is wise to avoid rubber underlays under a seamed carpet. Felts of natural fibres tend to pack down and lose their resilience in time and a felt and latex underlay combines many advantages of both rubber and felt. There are felts of 100 per cent courtelle with good durability and resilience. The underlay should be the same size as the carpet and in the case of stairs, if it is not continuous, the underlay should be in the form of pads which go right over the edge or nosing of the stair. When the underlay has a canvas backing or one smooth surface this is laid uppermost.

Stair and fitted carpets need fixing in some way, whereas a carpet square is generally laid loose on the floor. Stair carpets must be held taut, and for a wide staircase rods are better than clips and may be used in conjunction with tackless grippers. Carpet squares and rugs can be backed with thin foam rubber or rubberized net to prevent them slipping.

Suitable methods for fixing carpets

These include the following:

Glued – the carpet may be stuck to the subfloor.

Tackless gripper – steel pins protrude from plywood or metal strips fixed to the floor or stairs, and hold the carpet in place.

Turn and tack – the edge of the carpet is turned under and tacks put through the double surface but the carpet edge is inclined to ruckle and lifting is difficult.

Sunken – the carpet is laid in a sunken area, the edges of which are covered with brass or wood as for sunken front door mats.

Pin and socket ⎫
Ring and peg ⎪ These methods are especially suitable where a
Press studs ⎬ carpet needs lifting frequently, eg banqueting
Touch and close fastener ⎭ rooms cleared for dancing.

Where part of a carpet sometimes needs lifting a heavy duty zip fastener or lacing can be used to fix it to the remaining carpet.

Protection of carpets

Carpets may be protected from damage by:

a suitable underlay,

firm fixing so that rucks are avoided,

jute or coconut fibre mats at front doors etc,

dust control mats which may be impregnated on contract or made of synthetic fibres which attract and hold dust and dirt from shoes when walked
on,

strips of canvas, rubber or brass placed over the edge or nosing of a stair,

druggets, which are lengths of canvas or plastic laid as a covering over the
carpet to take the dirt and wear,

'jogging' a stair carpet, ie moving it up or down slightly,

turning a carpet round, so avoiding traffic lanes,

a flexible furniture arrangement which also avoids traffic lanes,

the use of castors,

spraying against moths,

the repairing of frayed edges and worn parts,

cutting and never pulling the pile when it 'sprouts',

protecting a damp carpet from metal castors or legs of furniture,

keeping dry cleaning fluid away from a rubber backing.

Cleaning of carpets

Cleaning is necessary in order to:

remove dust, grit and other soiling,

remove stains as soon as possible,

prevent damage by moth and carpet beetles,

retain the original appearance of the carpet as long as possible.

Daily cleaning

Superficial dust and crumbs may be removed daily with a carpet sweeper but
grit and other soiling which has got into the pile must be removed by suction.
The longer the pile the more thorough the vacuum cleaning must be and, in
any case, the vacuum cleaner should pass over the surface two or three times.
Stains should be removed as soon as possible; if liquid blot up with clean, dry
absorbent paper, if semi-solid or greasy material scrape up. Wipe over the
area with a damp cloth.

For residual stains use a solution of synthetic detergent, a dry cleaning
solvent or the appropriate stain removal agent (trying it out on an unseen
part of the carpet first) (see page 143).

Special (weekly) cleaning

At frequent intervals the edges of the carpet require special attention with a
damp duster, a brush or a vacuum cleaner and the whole carpet needs a
thorough vacuum clean. Shag pile carpets are rarely vacuum cleaned
thoroughly enough; they require vacuum cleaning in all four directions and if
the pile is very long it needs to be raked afterwards to make it stand up.

Periodic cleaning

This involves deep cleaning of the carpet and after thorough vacuum cleaning one of the following methods may be used:

shampooing,
hot water extraction,
dry foam extraction.

When *shampooing,* the detergent solution is released from a tank on the rotary scrubbing machine and the surface of the carpet is lightly scrubbed, and then left to dry. Finally the loosened soiling and detergent residue is picked up by suction.
The following points should be observed:

the detergent used should be one which dries to a powder so that a sticky residue is not left on the carpet,
excess water should be avoided so that the backing of the carpet does not become too wet,
each stroke of the brushes should overlap the previous one,
no metal, eg castors or legs of furniture, should come in contact with the damp carpet,
walking and replacement of furniture on the damp carpet should be avoided.

Hot water extraction machines are expensive machines with no rotary action. Hot water and the detergent are 'shot' into the carpet with high pressure spray nozzles. The dirt is flushed to the surface and the soiled water is picked up by the suction action of the machine. (There is a tank for the hot detergent solution and another for the soiled water.) The advantage of this machine over the rotary shampooing machine is that drying time is cut to a minimum (see page 103).
In *dry foam extraction* the cylindrical brush type machine lays down a moist foam, brushes it in and immediately extracts the soil-laden foam with a built-in vacuum head located behind the power brush.

Advantages and disadvantages of carpets

The *advantages*:

carpets add to the decorative appearance of the room,
they can give a luxurious appearance and feel,
they are warm and may keep out draughts,
they are quiet and afford some sound insulation,
they are non-slip,
carpet squares and rugs break up a floor surface and can be turned around,
fitted carpets make a room look larger, and there is only one floor surface to clean,
patterned carpets do not show stains as much as plain ones.

The *disadvantages*:

the surface holds dirt, so careful cleaning is necessary,
they are cut by grit and sharp castors and some are burned or melted by
cigarette ends,
they may be attacked by moth if made of wool,
shading can occur,
fitted carpets show definite areas of wear,
plain carpets show stains readily and stains can be difficult to remove
in situ,
a large pattern for a fitted carpet is extravagant because of the need to
match the pattern,
initial cost of good carpet is high.

Finishes

These can be:

Fire retardant, particularly when the carpet contains cotton or viscose.
Moth proof.
Water/stain resistant finishes, which are surface treatments of water
resistant silicone or fluoro-carbon products that have little effect on the texture or the appearance of the carpets, but provide a barrier between the fibre
and the spillage. Similar treatments may be given to upholstery coverings
(see p. 216).
Anti-static finishes contain substances which attract moisture from the
atmosphere so that the relative humidity immediately above the carpet is
high and the electricity leaks away. Anti-static finishes may include anti-
soiling properties. Anti-static fibres of stainless steel or copper may be introduced into the fibre or pile of a carpet to reduce the risk of electric shocks.
Disinfectant finishes may be applied during the manufacture or shampooing of a carpet. The spread of micro-organisms may be prevented by daily
spray vacuum cleaning when, by means of a suitable spraying device, the disinfectant fluid is sprayed into the pile of the carpet.

Mats and matting

Matting can be made from a variety of materials and the appearance and
wearing quality depends on the materials from which it is manufactured.
Coconut fibre, sisal, wood fibre, jute and rush are woven or plaited to form
matting. In addition, strips of plastic can be used. In all cases, because of the
loose weave, dirt filters through the matting to the floor below, thus
necessitating the frequent lifting of the mats, so that the dirt may be removed.
It is estimated that 80% of dirt is brought into buildings by the feet and the
use of mats and matting in entrances, foyers and corridors etc, can offer considerable saving in cleaning and maintenance costs. They help to prolong the
resistance to wear and to retain the appearance of the flooring.
There is a variety of materials used for mats in matwells and these include:

aluminium/rubber, which is prestigious and effective but somewhat
expensive,

coir, which is the traditional entrance mat and has excellent shoe drying properties. It should have a firm edge as this is the place which wears. The PVC-backed type can be cut to any size.

rubber link and *ribbed rubber,* which are extremely hardwearing and have effective scraper action,

ribbed polypropylene, which is a new development with high durability and ease of cleaning. There is a fast colour variety.

Dust control mats are used on floorings and carpets (ie not in matwells) and have a nylon pile to which dust and dirt are attracted by static electricity. The provision of clean dust control mats may be a contract service.

Vinyl transparent carpet protector runners are replacing druggets and have the advantage of showing the colour and pattern of the carpet.

Care and cleaning of mats

- Attend to worn or frayed edges.
- Coir mats – lift, turn upside down and tread on it to force the dirt out. Remove dust and dirt from the matwell before replacing mat. Vacuum clean.
- Plastic and other mats or matting – vacuum clean. Lift and vacuum clean underneath. Wash or scrub if necessary.

15

WALL COVERINGS

Wall coverings may be purely decorative, in which case the ability to bring colour, pattern, texture, light or shade to the room may be of the greatest importance; on the other hand the covering may be required to give an easily cleaned and hygienic surface. The choice is very wide and the style should suit the purpose, the furnishings and the architectural aspects of the room; however in all cases the wall covering must comply with the fire regulations. A large expanse of wall is noticeable and often there is less broken surface than on the floor, so wall coverings should blend with, rather than dominate, the general scheme. The need for warmer and more intimate interiors has led to the increasing use of textured wall coverings (see page 263).

It is possible to introduce more than one type of wall covering into a room and colours, designs or materials can be used for focal points to add interest. Areas where there is likely to be the greatest wear and tear can have the most durable surface and a choice can then be made on account of resistance to abrasion, tearing etc, as well as frequent cleaning.

Wall coverings inevitably become very rubbed and scratched by the movement of chairs, the carrying of luggage, the banging of trolleys and the rubbing caused by people as they pass. It is possible to guard against the marking of the wall in various ways, for example by:

  using a stronger and more easily cleaned material for the lower part of the
     wall, possibly up to 150 cm, which is called a dado,
  fixing narrow strips of wood to the floor or even to the wall, in such a way
     that chairs are not pushed right against the wall,
  using glass, perspex or melamine plastics as a protective material in
     vulnerable places, such as around light switches or walls against which
     people lean or which trolleys damage,
  fixing doorstops.

Generally, wall coverings may be cleaned by sweeping with a wall broom, vacuum cleaning with an appropriate attachment or washing with a sponge, warm water and synthetic detergent and, finally, rinsing.

Wall coverings should be used on suitably prepared walls, which often involves quite a lot of work and this adds considerably to the labour costs as in some cases the preparation entails more labour than the actual application of the new covering. The curing of the cause of any dampness, the filling of cracks and holes, the removal of dirt and the build-up of old paint and old papers are but a few of the jobs involved in preparation.

Types of wall covering

Paints

Paint is used extensively as a decorative wall finish, but it is also used to preserve and protect structural surfaces, especially those of wood and metal; it can be used for identification of pipes, for the emphasis of hazards and danger points, and for hygiene, as paint facilitates the cleaning of surfaces.

As a wall covering paint offers a wide choice of types, colours and degrees of gloss, and even design if murals are painted on to the walls. Whatever covering is chosen for the walls, gloss paint is normally used as a protective coating for window frames and sills, doors and skirting boards and in many instances it gives a contrast in colour and texture to the main wall finish, so contributing to the décor of the room.

Paint used for a wall finish is normally required for decoration rather than protection, but it also needs to be washable as paints tend to attract dust. (Stretch level dusting can in time become very noticeable and unless walls are washed from bottom to top, drip marks result on the dry surface and these are difficult to remove later.)

Paint is relatively cheap, easily applied and cleaned; it can give textural and multicoloured effects but it shows soiling (especially matt paints) and wall imperfections (especially gloss paints) more readily than other wall coverings. Although there are quick drying and low smelling paints, drying time and the lingering odour must be remembered when considering the time a room has to be 'off' for redecoration. There are paints with special properties, eg insecticidal, fire retardant etc, and these may be used to advantage in certain places.

The main types of paints used as wall finishes are: emulsion; alkyd; multi-colour; texture or 'plastic'; and microporous.

Emulsion paints are water-thinned but are based on dispersions of synthetic resins (eg polyvinyl acetate) which dry to tough, washable and wear-resistant films. Emulsion paints are available in a wide range of colours and various degrees of sheen from matt to semi-gloss or silk finish. Being alkali-resistant they are suitable for use on new, possibly damp, plaster etc. They are quick drying and low in odour, and so are very suitable for the redecoration of rooms which cannot remain long out of use. Silk finishes are not recommended for bumpy walls or ceilings where a high level of light reflection can be unpleasant, but look good over walls lined with plain or textured paper.

Alkyd paints are based on synthetic (alkyd) resins combined with a vegetable oil, such as linseed oil. They have almost completely replaced the older types based on natural resins, although the latter are still used in primers and undercoating paints. Alkyd paints are generally easier to apply and have better durability and wearing properties than the older types. They have good opacity and excellent light fastness in a wide range of colours. Polyurethane and silicones are sometimes included to give a more scratch-resistant surface.

Alkyd paints are available as gloss, silk and flat finishes. Some types are supplied as non-drip or 'jelly' paints which permit heavy application for maximum obliteration. Oil-based silk paints are good for walls subjected to heavy wear and condensation.

Multi-colour paints are usually dispersions of cellulosic colours in water. Each colour is present in separate 'blobs' or 'spots', the resulting effect being dependent on the number of different colours, the degree of contrast between them, and the size and distribution of the 'spots'. Usually this type of paint must be spray-applied. It is extremely hardwearing and the multi-colour effect helps to mask surface irregularities and imperfections. Corridors, entrance halls, hospital wards, cloakrooms and lavatories are ideal places in which to use this type of finish.

Texture or *'plastic' paints* are usually plaster-based and are intended to give a textured or relief effect on the surface. The texture is obtained by working over the material after application and, while it is still wet, using combs, palette knives, strippers, etc and much depends on the skill and taste of the operative. Some types are self-coloured, others may require painting when they are dry.

A modern type of texture material is based on a heavy-bodied synthetic resin emulsion and may be applied by spray direct to concrete and similar surfaces, thus eliminating the need for plastering. Such coatings are very tough and hard wearing, and their principal usage is similar to that of the multi-colour finishes.

Microporous paints have a rubberized base which gives little gloss but offers elasticity, allowing movement when the surface expands or contracts.

Care and cleaning of painted surfaces

1 Remove light dust with a wall broom or suction cleaner, working from the bottom up as dust tends to cling to the wall at downward angles.
2 Damp wipe or wash when necessary, with warm water and suitable detergent to remove heavily ingrained or tenacious dust and dirt. This is important on low sheen surfaces as dry cleaning tends to force dust into the surface.
3 When washing, start from the bottom and work upwards, using a sponge or distemper brush. Change the solution frequently. Rinse from the top downwards, using frequent changes of water. Sponge dry. Washing down is normally a maintenance job. There are wall washing machines for large areas, eg in hospitals, and the solution is then released under pressure to hand-held tools.
4 Low sheen finishes, especially emulsion paints, may tend to 'polish up' if isolated areas of bad soiling are rubbed vigorously with a damp cloth. Clean such areas by very lightly scrubbing with a damp nail brush and a little fine scouring powder; the dirt should be removed without damage or 'polishing'.
5 Never apply wax polishes or oil to gloss painted surfaces to 'revive' them. The residues may cause subsequent paint coatings to peel, or fail to dry.
6 Do not use harsh abrasives, strong solvents or strong soda solutions to clean paintwork, or the film may be damaged or softened.

Wallpaper

Most wallpapers in the UK are made in rolls of 10 m × 53 cm, but many foreign ones vary in length and width. The price varies enormously, depending on the quality of the design and the materials used.

Wallpaper may be smooth or have a textured effect introduced. This may be done by the superimposing or interlayering of other substances to give a rough surface, or by clever designing when apparent depth (three-dimensional effect) may result. Smooth finishes are more resistant to dust than rough surfaces, but marks generally show more.

The pattern may be of many kinds: floral, geometric, abstract, striped etc, of two or more colours, and in many cases the pattern is an all-over design. The choice should depend on the aspect, height, size and use of the room. Vertical, horizontal, receding and advancing designs and colours can all affect the apparent architectural features of the room. Large patterned papers tend to overpower and make a small room appear smaller; they also cut to waste owing to the need to match the pattern.

Wallpapers have a warmer appearance than paint,
- are not normally applied to new walls,
- offer some sound insulation,
- become soiled, scratched and torn with abrasion,
- can have torn and peeling pieces stuck back,
- may be given protection in vulnerable places.

Patterned and textured papers cover blemishes on the wall,
- cause problems when the walls are not 'true',
- without careful use can be disturbing.

Large patterned papers tend to overpower and make a small room look smaller,
- cut to waste owing to the need to match patterns.

New walls are not normally papered at first as it is wiser to allow them to dry out completely. In order to withstand steamy atmospheres and be spongeable, suitable papers may be treated during manufacture or after application to the wall, thus making wallpaper suitable for almost any type of room except perhaps large kitchens, laundries and hospitals.

In addition to the conventional wallpapers there are now available many paper-backed materials, eg fabrics, wood veneers and plastics etc, which are hung in a similar way to the conventional papers. Some of these newer coverings are extremely resistant to scratching, tearing and cleaning, and may be considered as being too durable and too expensive for areas where the décor is likely to be changed frequently.

The main types of wallpaper are:

ordinary surface printed papers;

spongeable papers, specially treated during manufacture to withstand water;

anaglypta, which has an embossed or raised pattern. It is white and normally painted over except when used as a ceiling paper;

wood chip papers, which have interlayered chips of wood, are usually cream in colour and are normally painted over;

oatmeal papers, where texture is produced by the interlayering of wood dust, chopped straw or similar material during manufacture;

flock papers, which are treated with adhesive to which silk, wool, cotton or synthetic fibres (the flocking) stick to give a raised pile;

wood grain papers, photographic reproductions of various wood grains, waxed during manufacture;

metallic papers, printed with gold and other metallic powders;

paper-backed hessians, which give a rough textured effect and are available in a large number of colours;

paper-backed felts;

paper-backed woven grasses or similar materials in beautiful natural colours;

paper-backed wools, with fine or coarse strands of wool in natural colours or bright dyes laid parallel fashion on a paper backing. They give a warm effect and improve insulation;

other *paper-backed materials*, including silks, linen, suede, and veneers of cork and wood;

lincrusta, a paper-backed textured composition, frequently simulating wood panelling.

Cleaning of wallpapers

1 Remove surface dust with a wall broom or suction cleaner (low suction for flock papers).
2 Remove marks by rubbing with a soft india rubber or a piece of soft bread. If the paper is spongeable, wipe with a damp cloth or sponge.
3 Attempt to remove grease with a proprietary grease absorber, although it may prove difficult to do so satisfactorily.

Plastic wall coverings

Many types of plastic wall coverings are available. Some are more decorative than others and some afford sound insulation but all, owing to their abrasion resistance, are more hardwearing and more easily cleaned than most other wall coverings. They are obtainable in a variety of sizes, with a great price range, and many require special adhesives. As they are non-porous there is a greater tendency for the growth of moulds so the adhesive should contain fungicides, or a fungicidal wash should be used on the wall prior to hanging the plastic wall covering.

The main types of plastic wall coverings are:

paper-backed vinyls, where the vinyl may have the appearance of almost any material, eg silk, tweed, hessian, cork, grass paper, wood, stone or brick,

fabric-backed vinyls, similar in appearance to the above and even more durable,

vinyl flock papers,

plastic wall tiles, imitating ceramic tiles,

laminated plastic, as a veneer or surface boards. Melamine is the resin frequently used during the manufacture of these plastic laminates which may simulate wood panelling or fabrics, eg Formica, Warerite, etc,

expanded polystyrene, in sheet or tile form, used on walls and ceilings to give heat and sound insulation, and to help eliminate condensation. It can be painted with emulsion paint or covered with paper. It is dissolved by spirit-based paints and thus before the use of oil paint it should be lined with paper and given a coat of emulsion paint to act as a buffer. There is a fire risk unless the polystyrene is treated.

Cleaning of plastic wall coverings

1 Remove surface dust with wall broom or suction cleaner.
2 Damp wipe or wash, when necessary, with warm water and synthetic detergent. A soft brush may be used on these surfaces.

Fabric wall coverings

It is possible to cover a wall with any fabric and its durability will depend on the fibre and weave used in its manufacture.

Fabrics may:

be hung loosely or in folds which may cover ugly features,
be attached to a frame secured to the wall,
be paper backed or specially prepared so that they can actually be stuck to the wall,
bring warmth and better acoustic properties to the area,
have sound deadening properties which help against noise in adjoining rooms.

Fabrics chosen should not be liable to sag, buckle or stretch when hung permanently on the wall and should not collect excessive dust or dirt. Dust and the smell of smoke tend to cling to fabrics with a pile or rough surface more than to smooth fabrics.

Wild silk and other beautiful fabrics may be padded for heat and sound insulation but silks and tapestries are expensive wall coverings and thus are more usually found in luxury establishments while hessian, linen and some acetate/viscose fabrics are cheaper and used more extensively.

The word 'tapestry' is frequently misused and applied to cross stitch (gros point) work on chair seats and stools, whereas true tapestry is a woven fabric and when used as a wall covering generally depicts a scene and hangs loosely on the wall, eg the tapestry in Coventry Cathedral. It should be remembered that wool materials may be attacked by moth and adequate precautions taken.

Cleaning of fabric wall coverings

1 Remove surface dust with a brush or suction cleaner.
2 For the more beautiful hangings, dismantle when necessary and send to a firm of dry cleaners who specialize in this type of work.
3 Where hessian is stuck to the wall, scrub very lightly where necessary using warm water and synthetic detergent.

Wood panelling

Woods used for panelling are usually hard, well seasoned and of a decorative appearance, and they may cover the wall completely or form a dado. Wood panelling may be solid or veneered; it lasts for years with little maintenance providing precautions are taken in respect of dry rot and woodworm, but the initial cost is high. Wood veneers may be stuck to paper, giving a similar effect to the solid wood at much less cost and veneered plywood panelling is also available. Wood panelling may be found in such places as entrance halls and staircases, assembly halls, boardrooms and restaurants.

Cleaning of wood panelling

1 Remove surface dust with a wall broom, duster or suction cleaner and polish if necessary.
2 Where the panelling has become dirty or greasy wipe over with white spirit, or vinegar and water, and repolish.
3 Dark oak may be wiped over with beer.

Glass wall coverings

Glass can be used in the form of decorative tiles, sometimes in the form of mosaics, and the tiles should not be confused with glass bricks which allow the passage of light and form the wall itself. Coloured opaque glass sheets or tiles may be used as a wall covering in hotel bathrooms.

Glass as a wall covering is frequently used in the form of mirror tiles which reflect light and can alter the apparent size of a room or corridor. Sometimes 'antique' mirror tiles are used, giving a duller surface with less reflection. Large unframed mirrors may cover part of a wall, eg over a vanitory unit or dressing table, while large framed mirrors are sometimes found on the walls of corridors etc.

A glass-less mirror is available now which has the advantage of not misting up or shattering and is about one-fifth of the weight of a conventional mirror. It consists of a polyester film, vacuum coated with aluminium and mounted on a flat frame.

Cleaning of glass wall coverings

1 Dust or wipe with a damp chamois leather or scrim. Proprietary cleansers or methylated spirit may be used.
2 Care should be taken when cleaning mirrors that the backs do not become damp.

Metal wall coverings

Metals may be used for their decorative and their hygienic qualities. Metals such as copper and anodized aluminium are decorative and may be used for effect in such areas as bars, where the metal in combination with rows of bottles and interesting lighting can be most impressive. Other metals, usually stainless steel in the form of tiles, may be used in kitchens where they present a durable, easily cleaned hygienic surface in areas where splashing is likely.

Metal skirting boards provide coved edges between wall and floor surfaces. Metal foil can be elegant if used sparingly as a wall covering; it is available in a variety of colours.

Cleaning of metal wall coverings

1 Dust or wipe with a damp cloth.
2 Polish is not necessary on metal surfaces as they either do not tarnish or will have been treated against it.

Leather (hide) wall coverings

Leather wall coverings are extremely expensive and very decorative. They may be padded and studded with brass studs and they do not usually cover a complete wall surface. They may be found in luxury establishments in parts of the restaurants or bars, but are too expensive to be found in most places and, in these, the effects of leather where required may be simulated by plastics.

Cleaning of leather wall coverings

1 Remove surface dust by dusting or careful suction cleaning.
2 Apply polish sparingly and rub up very well.

Other materials

Many flooring materials can be used as wall coverings. They contribute different colours, patterns and textures, depending on the particular material. They are usually hardwearing, resistant to abrasion and initially rather expensive. Some of them are particularly hygienic, easily cleaned surfaces and found in such places as kitchens, cloakrooms and bathrooms.

Amongst the floorings used as wall coverings are:

linoleum
cork
carpet
marble
terrazzo
ceramic tiles.

Ceilings

Colour, pattern and texture can be introduced into ceilings. Like walls, ceilings have the ability to affect the space, light, heat and acoustic properties as well as the appearance of the room. It stands to reason that the manner in which the ceiling is treated should be in harmony with the general décor.

The original ceiling of a room is generally plastered and almost any material may be put on to this. The ceiling may be papered, painted, decorated with wood in the form of beams or close slats, covered with tiles, eg acoustic, insulating or glass (including mirror), and in some special areas highly decorative mosaics are found, as are other materials, eg nails used for decoration.

In some cases use can be made of louvred woods, bamboo canes or metal grilles, for example, and in other cases grass cloths and other papers and fabrics may be used.

In some rooms suspended ceilings are useful and it is possible to introduce two-level ceilings in one room. Suspended ceilings are normally sheets of material supported on some form of framework and they offer a good opportunity for decorative effects, although their purpose may be to hide ugly details, eg lighting and ventilating fittings, pipes, etc. They may give better proportions in the room or emphasize a particular area, provide better acoustics and form an interesting lighting effect, as well as being purely decorative.

Dust and cobwebs should be removed from ceilings with a wall (ceiling) broom and for high levels of hygiene ceilings should also be washed.

16
FABRICS

Fabrics are used in a great variety of ways throughout the establishment. They may be chosen for their:

decorative value,
comfort,
warmth or coolness,
protective qualities,
durability,
and for hygienic reasons.

As the purpose for which fabrics are chosen varies, so does the wear and tear put on them; different fabrics may be subjected to:

soiling,
abrasion,
snagging,
creasing,
fading.

In order to obtain optimum use, fabrics should be chosen only after careful consideration, with a view to the purpose for which they are required.

In general fabrics are made by weaving yarn or threads spun from fibres which may be natural or man-made. There are fabrics in which threads are knitted together, and others are produced directly from fibres by fibre bonding or similar techniques, but textiles (woven fabrics) are used most commonly.

The physical and chemical properties of the fibres will contribute to the nature of the fabric, eg its:

softness,
durability,
elasticity,
lustre,
resistance to fading, soiling etc.

The properties of the fibres will also determine any treatment which may be given to the yarn or the fabric itself, eg dyeing, crease and shrink resistance etc.

As new fibres and treatments become available so the problems of fading, creasing and wear, and general maintenance become less troublesome.

The same or different fibres may be spun together and the resulting yarns or threads may be fine, thick, fancy, smooth or hairy. Any of these yarns may be used in open, close, plain, figured or pile weaves. In this way the appearance and characteristics of the final fabric are influenced by the type of fibre and the spinning and weaving processes used.

In the table opposite, examples of some of the names of the man-made fibres have been given but it is by no means a comprehensive list and it is the many trade names which can make these fibres so confusing.

Natural fibres

The most frequently used natural fibres are cotton and linen, which are of vegetable origin, and silk and wool, which are of animal origin. Vegetable fibres are made of cellulose and are similar in physical appearance and structure, whereas animal fibres are made of protein and there is a great variation in the physical form and structure of the different fibres (see table opposite).

Vegetable fibres

See table below for composition and characteristics.

Cotton

This is the most frequently used natural fibre; it is obtained from the seed of the cotton plant. The length of the fibre varies from 1.5 cm to 5 cm, the longest fibres coming from the West Indian and Egyptian cotton plants and the shortest from American and Indian plants. West Indian cotton, when spun and woven, produces a very fine fabric, Sea Island cotton, which is frequently used for dress materials. Egyptian cotton is used for tight, fine yarns suitable for the highest grade furnishing fabrics and for bed linen, while American and Indian cottons are used for furnishing fabrics of ordinary quality. The very short cotton fibres are called cotton linters and these are not spun into yarn, but used in cotton felt for bedding and upholstery, or in the

Characteristics of natural fibres

Vegetable fibres	Animal fibres
strong, with a crisp feel	soft
absorbent	absorbent
good conductors of heat	poor conductors of heat
not resilient or elastic	resilient and have elasticity and so resist crushing
stronger wet than dry	weaker wet than dry
affected by mildew if left in a damp condition	damaged by heat and disintegrate in sunlight
mothproof	damaged by alkalis and chlorine bleaches
not harmed by alkalis but lose strength in contact with acids	

Types of fibres

Origin of fibres			Trade names	Used in fabrics for
Natural fibres	vegetable	cotton		bed and table linen
		linen		soft furnishings, upholstery
	animal	wool		carpets, blankets, soft furnishings, upholstery
		silk		curtains, wall covering (in luxury establishments)
Man-made fibres	regenerated	viscose rayon	Viloft	
		modified rayon	Evlan, Sarille, Colvera,	soft furnishings
		acetate rayon	Dicel, Estron	carpets
		triacetate	Tricel, Arnel	
	synthetic	polyamide	Nylon, Enkalon, Celon, Bri-Nova	bedlinen, blankets, carpets, soft furnishings
		polyester	Terylene, Dacron, Trevira	bed and table linen, net curtains, fillings for pillows and quilts
		acrylic	Courtelle, Dralon, Acrilan, Orlon	blankets, carpets, upholstery, soft furnishings
		modacrylic	Teklan, Dynel, Verel	soft furnishings, blankets, upholstery
		polyvinyl	Saran, Movil	certain types of upholstery and deck chair coverings
		polyethylene	Courlene, Polital	upholstery
		polypropylene	Curnova, Spunstron	carpets
		glass fibres	fibreglass	curtains, fireblankets

production of rayon.

Cotton fibres are flat and ribbonlike, and have a natural twist which aids the spinning and makes a strong thread (ie it has good tensile strength).

Although good conductors of heat, cotton fibres have a hairiness which will trap air. This means that cotton materials have a slight feeling of warmth, which may be increased when the material is brushed or teased since more air is held. The hairiness of the fibre accounts for cotton materials being unsuitable for polishing glass, as linters or 'bits' are left on the surface; it also means that cotton materials soil more easily than ones made from smoother fibres, eg linen.

Cotton has little resilience so creases occur in cotton fabrics; and if cotton yarn is used in the pile of carpets it soon has a flattened appearance.

Mercerized cotton is produced by treating the yarn or fabric with caustic soda, under conditions where the fibres are stretched. Mercerization results in smoother, more rodlike fibres which have a gloss or sheen, and it gives cotton a greater affinity for dyes and improves its strength.

Linen

Linen is obtained from the stem of the flax plant. The length of the fibre, which can be from 500 cm to 1 m long, enables a fine, strong yarn to be spun.

The fibres are smooth, straight and almost solid, and these factors account for the chief differences between cotton and linen fabrics. The smoothness of the fibres gives good dirt and abrasion resistance, so linen fabrics are popular for loose covers; it also accounts for the lack of linters, making linen suitable for glass cloths, and for the fabric being cool to the touch. The smoothness and the straightness of the fibres give linen its lustre which can be increased by mercerization when its strength and dye affinity are also increased. The solidity of the fibres makes linen much heavier than cotton and the weight of a pile of linen table cloths or sheets on a shelf can be considerable.

Linen has little resilience and creases badly, the creases being much more sharply defined than in cotton fabrics – linen sheets become crumpled very much sooner than cotton ones.

Thread spun from the shorter linen fibres is known as tow yarn, and this yarn produces a softer, less 'rigid' and more absorbent material than that spun from the longer fibres, line yarn; tow yarn is therefore more suitable for towels, glass cloths, and fabrics which need to drape well. Line yarn produces a strong material with more resistance to dirt and abrasion, and is used for bed and table linen and upholstery.

Linen is the only fibre name which also applies to a fabric.

Jute, ramie, hemp and sisal

These are also natural fibres of vegetable origin, the first three coming from the stems of plants and the last from the leaves of a plant, but they are not used in fabrics to the same extent as cotton and linen.

Jute, hemp and sisal are used for twines and sacks; jute is also used for hessian, and the backing of carpets and linoleum; sisal is used for the manufacture of mats. Although the main use of hemp is for twines, canvas and sacks, it can be made into lustrous fabrics. Ramie is very strong with a fine natural lustre and was used in high grade furnishing fabrics, especially pile fabrics, but because of various difficulties, mainly in supply, it is not used so much now.

Kapok

Kapok is obtained from the seed of a type of cotton tree. The fibre is smooth, light and lustrous. It is used for the filling of cushions.

Animal fibres

See table on page 202 for composition and characteristics.

Wool

Wool normally means the fibre from the fleece of the sheep, but fibres from other animals, eg horse, camel, llama and goats, are also used. Most wool is the yearly growth from the living animal, that is, fleece or virgin wool, but 'skin' or 'pulled' wool may be obtained from the bodies of dead sheep, and remanufactured wools may be obtained from used wool. Of the remanufactured wools, 'shoddy' is perhaps the best for re-use. The wool fibres are pulled out as knitted woollens etc are run through garnetting machines and the fibres may then be respun and made up into cheaper woollen articles.

Wool fibres vary in length from about 4–40 cm and they also vary in diameter, some being very much finer than others. They have a natural crimp or wave, which gives wool its elasticity and resilience, enabling it to resist crushing, and making it particularly suitable for carpets and upholstery. The fibres are not smooth but have overlapping scales which enable air to be held between the fibres, and causes woollen material to feel warm.

These scales interlock with friction, as for example in careless laundering, and bring about felting and shrinkage when the material will be less warm. In cut pile carpets this felting is an advantage as the loose fibres, produced as a result of the cutting of the pile yarn, become bedded down in the carpet, provided little vacuum cleaning is carried out when the carpet is new.

Wool fibres are used to give woollen or worsted yarns. In woollen yarns, the fibres lie in all directions and result in a hairy fabric, but for worsted yarns the fibres are combed parallel and the fabric produced from these smoother, more tightly twisted threads, is less rough and hairy, and is more expensive.

Wool cannot be dyed to the same standards of uniformity or fastness as vegetable and man-made fibres, and unlike these it is attacked by moth. Wool can be mothproofed, but care must be taken to see that the proofing withstands any cleaning process.

Figure 16.1　Woollen fibre

Silk

Silk is obtained from the cocoon spun by the cultivated silkworm, in the form of long filaments which may be on average 274–456 m long. The cocoon may consist of up to 3 km of filament, but this cannot be unwound continuously.

The filaments are smooth and tubelike, with no irregularities and the beautiful lustre of silk fabrics is due to these properties. Silk is stronger than cotton (ie it has greater tensile strength); it is elastic and resilient and so it does not crush easily. It is, however, weakened when wet, and it disintegrates in sunlight.

Spun silk is manufactured from the shorter filaments obtained during the 'reeling' or unwinding of the cocoons, or from pierced cocoons, and it is less smooth and lustrous than thrown silk made from the finest filaments.

Silkworms of the wild silk moth, such as the Tussah moth, produce wild silk, in which the filaments are frequently irregular, producing slubs or variations in thickness when the silk is woven. The natural gum is not completely removed as is the case in silk from the cultivated silkworm.

All silk materials have elegance but due to their expense their use is normally confined to luxury establishments.

Asbestos

In addition to natural fibres of vegetable and animal origin there is a mineral fibre – asbestos, which occurs naturally in many parts of the world. It is quite incombustible and was used for fireproof materials. It is now known to be a safety hazard and glass fibre is used instead for fire blankets etc.

Man-made fibres

The number of man-made fibres has increased tremendously in recent years and there can be no doubt that still more and improved fibres will be discovered as research continues. All man-made fibres are made in basically the same way. The raw materials are treated chemically to form a viscous liquid, after which long continuous filaments are produced, in varying thicknesses depending on the size of the spinning jet used. The long filaments may be cut up into fibres of required lengths and yarns from these cut-up or staple fibres produce a less shiny fabric, with a softer and warmer handle than the continuous filament yarns. Some staple fibres may be 'bulked' to give an even warmer handle and improved absorbency. Man-made fibres are mothproof and the majority are mildew proof and have a low moisture absorbency.

Man-made fibres may be *regenerated*, eg rayon, when the fibres are retrieved from natural substances, the most usual being cellulose, or *synthetic*, eg nylon, when the fibres are produced by chemical synthesis (ie they are built up from basic chemicals).

Regenerated fibres

Rayon

The term 'rayon' (see table, page 207) often covers viscose, acetate and cuprammonium fibres but while viscose and cuprammonium rayons have

Characteristics of rayon

Viscose rayon	Acetate rayon	Triacetate rayon
absorbent, little resilience, flattens and creases, smooth filaments so has lustre, weaker wet than dry and fabrics tend to lose shape, dyes well, harmed by bleaches and other chemicals, decomposes at 185–222°C without melting, fibres bonded together for 'disposables', frequently blended with fibres of greater strength eg cotton, and fabric then used for upholstery, loose covers and heavy curtains, not suitable for furnishing fabrics requiring frequent laundering.	not so absorbent, little resilience, softer handle, drapes well, more silk-like, poor wet strength, dissolved by acetone, harmed by acids and alkalis, softened at 180°C and melts at 230°C, fabrics only suitable where not subjected to great strain eg lightweight curtains and bedspreads, not suitable for furnishing fabrics requiring frequent laundering.	low moisture absorbency, and fabrics drip dry, more resilient and relatively crease resistant, has lustre, softened by acetone, harmed by alkalis and trichlorethylene (dry cleaning fluid), melts at 260°C, may be 'bulked' for filling of quilts, used for 'pile' of candlewick bedspreads.

many properties in common, they differ considerably from rayon acetate. It would, therefore, seem wiser to restrict the term rayon to viscose rayon, as is more usual in the USA. Cuprammonium rayon has not the importance of viscose and acetate rayons in the field of furnishing fabrics, and will not be considered here.

Viscose rayon filaments consist of pure regenerated cellulose obtained from wood pulp or cotton linters. *Rayon acetate* is also obtained from wood pulp or cotton linters, but as it is a cellulose derivative, and not pure cellulose, it has different properties from viscose rayon. The basic properties of viscose rayon can be modified during manufacture and there are now a number of modified rayons, eg Evlan, Sarille and Durafil.

Evlan has improved durability and resilience and was developed especially for carpets; it is used in blends for tufted and woven carpets as well as in 100% form. It is also used extensively in upholstery fabrics. Evlan M is an improved version and Evlan FR is flame resistant.

Sarille has a high degree of crimp and so produces fabrics with a wool-like handle. It has crease shedding properties and is used for blankets and candle-wick bedspreads, when it has the advantage of being lint-free.

Durafil, because of its improved durability and resistance to abrasion, can be used in blends for uniform and upholstery fabrics.

Regenerated protein fibres

There are, in addition to the regenerated cellulosic fibres, regenerated protein fibres. The fibres may be extracted from the proteins of milk, corn, ground nuts and soya beans but have relatively little importance compared with the other man-made fibres. Fibrolane is one obtained from casein and has wool-like properties. It is used to some extent in the manufacture of carpets.

Synthetic fibres

Synthetic fibres are grouped according to their chemical composition, for example:

polyamide fibres, eg nylon,
polyester fibres, eg terylene,
acrylic fibres, eg acrilan,
polyvinyl fibres, eg saran,

polyolefin fibres { polyethylene, eg courlene,
{ polypropylene, eg spunstron,

Their individual names, however, prove confusing as different manufacturers give their own trade names to fibres of similar properties; thus Perlon is the German brand of nylon, and Dacron the American polyester.

Most synthetic fibres are produced as continuous filaments which may be cut into staple form; these may, in some cases, be 'bulked' to give a softer and warmer handle. Bulked fibres tend to 'pill' (ie the nap forms tiny balls) especially in laundering, and once the nap has gone the article is not as warm.

Synthetic fibres (with the exception of glass fibre):

have great strength,
have resistance to abrasion, moth and mildew,
have resilience and elasticity, so fabrics do not crease easily,
have low moisture absorbency so fabrics drip dry,
are thermoplastic, so soften in hot water during laundering and absorb
grease,
are poor conductors of heat,
are not normally harmed by dilute chemicals,
are electrostatic and attract dirt. (However, epitropic fibres with antistatic

Polyamide and polyester fibres

Polyamide (nylon)	Polyester
produced in various forms, melting points vary, 185°–250°C, fabrics easy to wash and dry and should be washed frequently because dirt is attracted, great strength, elasticity and abrasion resistance makes it very suitable for upholstery fabrics and carpets, also used for bed linen, blankets and uniforms, can be brushed or 'bulked' so that more air is held and fabric warmer.	melts at 243°C, very low moisture absorbency which makes dyeing difficult, more electrostatic than nylon or acrylics, good resistance to abrasion but inferior to nylon, creases hold more easily than in nylon, greater resistance to sunlight so very suitable for net curtains, when blended with cotton or wool is used for bed and table linen and uniforms, may be 'bulked' and used for pillow and quilt fillings.

and electrically conductive properties have now been produced, eg Ultratron, an antistatic nylon)

For sheer strength nylon is the best of the synthetic fibres, but for ease of cleaning and resilience the acrylics are better.

Acrylic fibres

Acrylics are produced in various forms	

Acrilan
most wool-like of all synthetics,
has fluffy, soft handle,
good resilience and crease recovery,
melts at 240°C,
good resistance to chemicals and sunlight,
less resistance to abrasion than nylon but
  better than wool,
suitable for blankets, carpets and
  upholstery.
	Courtelle
	does not melt but sticks to iron at 154°C,
	readily takes permanent twist so
	  particularly suited for twisted pile
	  carpets and also used for blankets.

Dralon
similar to Acrilan and used for curtains,
  upholstery fabrics and in carpets.
	Orlon
	has excellent resistance to sunlight so very
	  suitable for curtains.

Teklan is a modacrylic fibre. It is flame retardant and used in carpets, curtains and upholstery fabrics.	

Saran

This is a *polyvinyl fibre*. It is almost completely non-absorbent and therefore difficult to dye. Saran does not burn but loses strength in boiling water and softens at 121°C. It is used for certain types of upholstery fabrics and deck-chair coverings.

Courlene

This is a *polyethylene fibre*. It is non-absorbent, and the fibres soften about 95°C (ie below the temperature of boiling water) and melt about 120°C. It is used for certain types of deck-chair coverings, awning and upholstery fabrics as well as plastic floor mattings.

Polypropylene

These fibres are the lightest in weight of the synthetic fibres; they have the greatest abrasion resistance and are used in some of the adhesively bonded felted carpets.

Glass fibre

This is produced in the form of fine filaments from molten glass. It is non-absorbent, highly resistant to chemicals and strong sunlight and it is fireproof (melting point 815°C).

Glass fibre has a poor resistance to abrasion; the fibres are brittle but when fine they can be woven into fabrics for lightweight curtains, and for shower curtains and fire blankets. In some instances it is reinforced with polyester resins, and produced as a laminated sheet material which has great strength and durability, and is used for baths, lavatory basins and sinks (see page 158).

Identification

The identification of textile fibres has become very complicated with the advent of so many man-made fibres and the increasing use of fibre mixtures, and the special treatments given to some fibres have added to the difficulties.

It is possible, however, to examine:

appearance, feel and length;
appearance under the microscope;
behaviour to heat and flame;
behaviour and solubility in certain acids, alkalis and organic solvents;
reaction to colour staining solutions, such as Shirlastain.

Fabrics

The threads spun from the fibres may be knitted or woven into a fabric or material. (Many disposable and short-life articles are made direct from fibres by felting or bonding.)

Figure 16.2   Knitting

During knitting, one thread is normally used to form a series or row of loops which in turn is held by another row and in this way stocking stitch and other stitches are formed. Stockinette swabs, certain types of loose covers (stretch) and some uniform materials are made in this way.

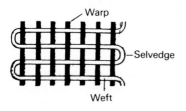

Figure 16.3   Showing the selvedge

Woven fabrics

Household fabrics are normally woven, that is, they are textiles. In weaving there must be two sets of threads. The loom is set up or 'strung' with long vertical threads, called the warp, and crossing or interlacing these threads is the weft. The weft thread is carried by a shuttle and turns at the side of the series of warp threads to form a firm edge – the selvedge. Many widths of material may be produced.

Yarns for warp or weft may be spun from two or more types of fibres and the fabric is then described as a blend; a mixture indicates that a fabric is manufactured from two or more different yarns, eg nylon warp and rayon or cotton weft. By mixing and blending a variety of textures and colour, different effects can be introduced into fabrics and they may be given improved properties, eg greater strength, better drape or washability etc.

The appearance, quality and behaviour of any fabric in use are dependent on the nature of the weave as well as on the type of fibre used in the yarns. Hence certain fabrics are more suitable for some purposes than others.

In the manufacture of textiles there are several standard weaves, eg plain, twill, satin, huckaback, damask, pile, etc (see Fig. 16.4 below).

Plain   Twill

Satin   Damask

Figure 16.4 Frequently used weaves

Plain weave

The weft goes over and under alternate warp threads, as in darning, in plain weave.

Fabrics made in a plain weave are normally smooth, and their firmness will depend on the number of warp and weft threads per cm, that is, on the closeness of the weave. A close weave gives a strong cloth which keeps its shape, but one which tears easily. The weight and appearance of plain weave fabrics vary enormously according to the thickness, character and closeness of the threads, none of which need be the same for warp and weft.

Chintz, cretonne, scrim and sheeting are fabrics made in a plain weave and it forms the base of many pile fabrics. Repp is a plain weave fabric where the ribbed effect has been produced by using warp and weft of differing thicknesses.

Twill weave

The weft threads cross the warp at different intervals in the different rows in a twill weave, so that a series of diagonal lines is produced on the surface of the fabrics.

Herringbone patterns are produced by reversing the direction of the twill at regular intervals across the width of the cloth. In a twill weave the threads are normally close together and the finished fabric is firm and hardwearing.

Twill sheeting, drill and gaberdine are fabrics made in a twill weave.

Satin weave

There are fewer intersections of the warp and weft threads in a satin weave and the intersections are uniformly distributed. The warp 'floats' over, for example, four weft threads and forms the surface of the fabric, ie the fabric is warp faced.

The fabric is smooth, usually of dense construction and with an attractive sheen. Owing to the 'floating' threads, it may present little resistance to abrasion and become pulled or snagged.

There are many types of furnishing satins, that is, furnishing fabrics made in a satin weave, eg cotton satin, rayon satin, etc.

A sateen weave is similar to a satin weave but it is the weft threads which 'float' and the fabric is therefore weft faced.

Figured weaves

These introduce a pattern into the fabric. The pattern may be introduced by combining two of the previously mentioned weaves, as in damask, when for table linen warp and weft satin weaves are combined, and for soft furnishings warp satin and plain weaves are combined; or the pattern may be introduced by the use of coloured threads additional to the foundation cloth, as in brocade and tapestry. Huckaback and brocatelle are other fabrics woven in a figured weave.

In all these cases, the weaves are produced on a Jacquard loom, where each warp thread is controlled individually; the thread as it is required is lifted by a harness, controlled by a series of punched cards corresponding to the pattern to be woven.

Fabrics made in figured weaves vary considerably in weight and appearance.

Pile weave

In a pile weave there are tufts or loops of yarn which stand up from the body of the cloth. The tufts or loops are extra warp or weft threads woven at right angles through the foundation cloth. These extra threads form the pile or surface thickness and they may be cut or uncut. Patterns can be produced by combining cut and uncut piles (cf carpets).

A thick, hard wearing fabric results which has a tendency to 'sprout', snag and collect dust.

Velvet is a cut, warp pile fabric; turkish (terry) towelling is an uncut pile fabric with the pile on both sides of the material; moquette may be either a cut or uncut pile fabric.

Pile fabrics may also be made by *tufting*, eg candlewick, when the pile is needled into an already woven foundation cloth or backing.

Cellular weave

These weaves give a loosely woven fabric which holds air in the 'cells' between the threads, eg cellular blankets.

Originally certain fibres were associated with particular fabrics but this is no longer the case and many fabrics are obtainable in the different natural and man-made fibres. For example, satin was originally always made from silk and made in a satin weave but now cotton, rayon and nylon satins are available.

Fire-resistant fabrics

Pure wool, glass fibre and modacrylic fabrics are inherently flame retardant. All cotton, linen and most rayon fabrics can have applied finishes which render them flame retardant. Flame-resistant fabrics should satisfy BS 3120 and BS 5867 (1980) Pt. 2 for use as curtains and upholstery coverings in public rooms in all establishments, as well as for bedspreads and cubicle curtains in hospitals.

Finishes given to fabrics

Water repellant (eg Dri-Sil, Velan)

This is a silicone finish applied to a fabric (including carpets) which causes the spillage, eg fruit juice, to 'pearl' off the fabric. It is durable to washing but cotton and acetate fabrics show less durability than others.

Oil repellant (eg Scotchgard)

This is a fluorochemical finish which gives both oil and water repellancy. Stains stand on the surface and can be blotted away. It is expensive but resists laundering and dry cleaning.

Soil release (eg Permalose)

This is an anti-greying finish which may be applied to fabrics containing polyester.

Crease resistant and easy care (eg Calpreta, Bel-o-fast)

These are resin finishes which prevent creases forming and are applied to cotton, cotton blended with synthetic fibres or viscose fabrics. This can result in loss of strength – tear, tensile and resistance to abrasion. Boiling, bleaching or excessive agitation during washing or spin drying may destroy the finish. (Continued on page 216.)

Characteristics and uses of fabrics

Fabric	Fibre	Weave	Appearance	Uses
Baize	all wool	plain	surface of cloth is raised and closely cropped, generally green colour	aprons, storing silver, covering meeting and boardroom tables
Brocade	silk, cotton, rayon or synthetic	figured	extra weft threads give colour and form pattern	upholstery, curtains, bedspreads
Candlewick	cotton, rayon, sarille, triacetate, nylon, courtelle	plain foundation cloth	tufted yarns are inserted into foundation cloth and give pattern	bedspreads, bathmats when made from absorbent fibres
Calico	cotton	plain	thicker than muslin, white, dyed or unbleached	sheeting, dust covers
Chintz	cotton	plain	printed pattern, glazed finish	curtains, loose covers, bedspreads
Corduroy	cotton	pile (cut)	cut pile forms lines or cord in warp direction, dyed	upholstery
Cretonne	cotton, rayon	plain	printed pattern, coarser than chintz	curtains, loose covers
Damask – table	linen	figured	warp and weft faced satin weaves give design and background, self-coloured	table linen
– furnishing	cotton, linen, silk, rayon, polyester and any combination	figured	warp satin and plain weaves give better differentiation of design, sometimes further emphasized by different fibres for warp and weft threads, self-coloured, great variety of weights	upholstery, curtains, loose covers
Denim	cotton	plain	warp dyed, weft undyed, speckled effect	overalls
Drill	cotton	plain	both warp and weft white or dyed	overalls
Felt	wool, courtelle		densely matted fibres	to protect tables, carpet underlays
Flannelette	cotton	plain	brushed (teased), fluffiness holds air and fabric is warmer than ordinary cotton fabric	sometimes for sheets, underblankets
Folkweave	cotton	loosely woven	coarse coloured yarns in simple designs, often with textured or three-dimensional effects	curtains, bedspreads

				covering
Hessian	jute	plain	lawn coloured but may be dyed	
Huckaback	linen, cotton, rayon	figured	huckaback weave	face, hand and continuous roller towels
Moquette	cotton, wool, rayon	pile	cut or uncut pile or combination of the two, wool pile on cotton ground is best quality but cotton and rayon piles are hard wearing	upholstery
Net	many fibres, polyester best	open plain	threads may be twisted or knotted instead of woven, white or coloured	glass (sheer) curtains
Plush	cotton, silk, man-made	pile (cut)	deeper pile than velvet but less closely woven, hard wearing	upholstery, curtains
Repp	cotton, silk, wool, synthetics	plain	fine warp and coarser weft gives rib running from selvedge to selvedge, hardwearing	upholstery, curtains, loose covers
Sateen	cotton	satin	weft faced, smooth	curtain lining, underside of quilts
Satin	silk, rayon, cotton, synthetics	satin	warp faced, smooth, lustrous, variety of weights, silk originally now cotton satin, rayon satin etc	curtains, bedspreads, cushion covers
Scrim	linen	plain	lint-free cloth	window cleaning etc
Tapestry	wool, cotton, rayon or mixtures	figured	pattern formed by extra coloured weft threads, closely woven	upholstery, curtains, loose covers
Ticking	cotton, rayon	twill or satin	closely woven striped or coloured fabric	mattress covers, white for enclosing filling for pillows or cushions
Towelling	cotton	pile (uncut)	reversible uncut pile, turkish (terry weave)	towels, bathmats
Tweed	wool	plain or twill	heavy fabric, plain or coloured	upholstery, curtains
Velour	cotton, rayon	pile (cut)	warp pile, has tendency to lie down	upholstery, curtains
Velvet	silk, cotton, rayon, synthetics.	pile (cut)	warp pile, great variety in weight	upholstery, curtains, cushions
Velveteen	cotton, rayon, mercerized cotton	pile (cut)	weft pile is generally shorter than velvet	upholstery, curtains, cushions
Wild silk	silk from silkworm of wild moth	generally plain	filaments are irregular in thickness and produce slubs in the fabric, expensive, hard wearing	curtains, wall covering

Shrink resistance (eg Sanforising, Rigmel)

The fabric is pre-shrunk for this finish.

Non-felting for wool

This is a film of synthetic fibre, eg nylon, applied to wool to reduce tangling, or the wool may be treated with chemicals to reduce the scaly part of the fibre. This renders the fabric shrink-proof and gives a high level of washability.

Mothproofing (eg Dielmoth, Eulan)

A substance poisonous to moth maggots is applied to the fabric or fibres to obtain a moth-proofed finish.

Bacteriotatic protection (eg Actifresh, Durafresh)

This is provided by a chemical which inhibits the growth of bacteria which decompose perspiration.

Flame retardant (eg Pyrovatex, Proban)

This finish may be applied mainly to cellulose fibres, eg cotton, linen, rayon. It is fast to laundering provided the temperature is less than 50°C and a soapless detergent is used. (With soap, particularly in hard water, a lime soap is formed which destroys the applied finish.)

Wool is generally considered flame retardant for most purposes but flame retardant treatments, eg Zippo, may be applied to woollen carpets and soft furnishings, eg curtains, chaircovers. Articles may be washed in synthetic detergent at 40°C or may be dry cleaned without losing the finish.

Certain synthetic fibres are inherently flame retardant, eg Teklan, Dynel.

There are firms which will test samples of fabrics and report on their suitability for fire resistant finishes.

BS 5867 Part 2 (1980)

This is the most important specification for flammability of fabrics.

'C' level of performance is recommended for hotels etc. The fabric is required to be tested twice, once in new condition and again after a cycle of 50 washes.

'B' level of performance is the usual requirement.

'A' level of performance is not accepted by the fire officer.

17

SOFT FURNISHINGS

Soft furnishings include curtains, loose covers, cushions, bedspreads and quilts (but not carpets), and they contribute greatly to the appearance of the room by bringing to it colour, pattern and texture. Some articles give protection and some, in addition, give warmth and comfort. As each is subjected to different types and amounts of wear and tear, it follows that the fabric from which it is made should be suitable for the purpose.

Curtains

Curtains are required to:

give privacy where windows may be overlooked,
darken the room when necessary,
reduce heat losses and noise levels,
bring character and atmosphere to the room by their line, colour, pattern and texture.

Curtains are essential to the appearance of any window and almost any furnishing fabric can be used. However, the weight, colour and pattern of the fabric will be determined by the size and position of the window, and the general character of the room.

The fabric should be chosen with regard to its resistance to fading and abrasion, its drape, dimensional stability and its flame resistance.

The fabric's life expectancy is related to the amount of its exposure to:

sunlight,
airborne soiling.

These cause fading and rotting of the fabrics but curtains may be lined, thus partly reducing the damage. Yellowing due to fibre oxidization occurs mainly in cotton, linen and rayon fabrics. Glass fibre, terylene, acrylics, saran/viscose rayon blends and brightly coloured nylon fabrics withstand sunlight well.

'Colourfast' can be an ambiguous term as it does not necessarily mean that the colour is fast under all conditions of use. In addition to fading by sunlight colour may be lost by:

water spotting,
washing,
dry cleaning,
abrasion.

Curtains are subjected to abrasion by:

being pulled (drawn),
being brushed against,
rubbing along the floor, window frames and sills,
being laundered or dry cleaned.

The abrasion resistance of the fabric will depend on the fibres from which it is made and the construction of the yarn and the fabric. Synthetic fibres, especially nylon and polyester, have excellent abrasion resistance and wool, cotton, linen and high tenacity rayons wear well in this respect also. Silk has most of the properties required for curtains but it is very expensive and silk fabrics (often wild silk) are normally only used in luxury establishments.

Curtains may be a fire hazard and the ease with which a fabric burns depends on the fibres from which it is made, the flame retardance of the fabric or the finish given to it. Pure wool, glass fibre, asbestos, saran blend and most acrylics, eg teklan, are inherently fire retardant and a fabric which is closely woven is less likely to burn than one which is loosely woven. Flame retardant fabrics satisfying BS 3120 are required for curtains in all public areas of an establishment and for cubicle curtains in patient bed areas in the National Health Service. They are desirable throughout establishments. If curtains are flame-proofed they should be marked as such because flame proofing may be affected by laundering and make reproofing necessary.

Curtains have a poor appearance if they do not hold their shape and drape well. Loosely woven materials tend to drop unevenly and with constant hand drawing the sides of the curtains may go out of shape.

Curtains need to be suspended from a horizontal rod or traverse track which may be of metal, wood or plastic and which may be the width of the window or extend either side of the window frame. When the latter is the case the curtains can be drawn well back, so:

giving width to a narrow window,
preventing the fading of the folds of material nearest the window,
allowing more light into the room (see page 264).

If curtains do not really meet in the centre of the track the gap left gives an untidy appearance from both inside and outside the window. This gap may be prevented by having an overlap in the centre of the track. This too may be unsightly when the curtains are drawn back, even with the modern curtain headings, unless it is hidden by a pelmet or valance. Some modern tracks or rods have a better appearance than others and decorative headings are usual in many places.

Curtains are fixed to the track by rings or hooks and drop to the floor or to the window sill where architectural aspects dictate, ie the shape of the window, radiators, or fitted furniture beneath it. They are normally of some opaque material, with sufficient fullness so that they can be drawn across the window to give complete privacy, some heat and sound insulation, and a good appearance.

Figure 17.1 Pinch pleated curtain suspended from rod and rings

Figure 17.2 Pinch pleated curtain suspended from modern rail

There is a wide variety of fabrics suitable for ordinary curtaining, and the choice will depend on the type of establishment and the particular room for which it is required. Sometimes different curtains are provided for winter and summer use.

When choosing fabric for curtains it should always be seen in a large piece, hanging in folds as the pattern quite frequently looks different when lying flat. It should be remembered that the light comes from behind the material during the day whereas at night it falls on the material. Materials with white backgrounds often lose their whiteness and fresh appearance after several cleanings, and so are best avoided. If a material has a large 'drop' in the pattern wastage may occur when matching up the curtains. Curtain materials may be 80 cm, 120 cm or 126 cm in width while some are much wider, and in order that the curtains hang well, the minimum width of the curtain should be 1½ times the width of the rod or track (see page 263).

Good curtains are usually lined and heavy curtains are often interlined. The lining enables the curtain to hang better, to be protected from dirt and sunlight, to give a uniform appearance from outside, and to afford greater insulation (ie keep heat in and cold out). The lining is often cream or beige sateen (or a colour which matches the background of the curtains) and is obtained in widths of 91 cm, 121 cm and 137 cm. A metallic fabric, milium, having good insulating properties, can also be used as a lining material with the flannelette fabric 'bump' used for the interlining.

For public rooms it is best to buy as good a material as can be afforded; such fabrics as brocades, damasks, heavy satins, wild silk, printed linens, tapestries, velvets, novelty weaves and many others are suitable in the right setting.

For bedrooms, a lighter material can usually be used and will probably be less expensive. Printed cottons and linens, chintz, damask, cotton and rayon satins, repps and cretonne are some which may be chosen.

In bathrooms the windows are often of opaque glass so that curtains are not necessary, except for appearance. If used, curtains are usually unlined and of some easily washable material and nowadays colourful blinds are sometimes used instead of curtains. Nylon, plastic and glass fibre materials are sometimes used to curtain the showers, as these materials are non-absorbent and easily sponged of splash marks, but plastic material needs careful pulling because it goes stiff in time and with careless use tears can occur.

Figure 17.3   Curtains used as room dividers

In some places curtains are found in rooms with no windows, eg lower ground floor banqueting rooms, and with careful use of fabric, lighting and air conditioning, the lack of windows may not be apparent. In some hotel bed-sitting rooms curtains are used to divide the bed area from the sitting area (see Fig. 17.3) and in hospitals curtains separate beds (cubicles).

In any one establishment there are many curtains of varying lengths, so it saves endless trouble if they are marked for the various rooms. Curtains should never hang long without cleaning as dust and soot rot the fabric. They may generally be laundered or dry cleaned but lined curtains are better dry cleaned, as in laundering the lining and curtain materials may shrink at different rates.

Care and cleaning of curtains

1 Keep rod and track free from dust by the use of a wall broom or a vacuum cleaner attachment.
2 Shake often. Use vacuum cleaner attachment or brush occasionally to remove dust from the curtains.
3 Deal appropriately with repairs to linings, frayed edges and any difficulties with pulling, bent tracks, etc.
4 Reverse the position of the curtains so that no part is continually exposed to the sun.
5 Have lined curtains dry cleaned; if unlined, and of a suitable material, have them laundered.
6 Solvent spray extraction machines may be used similar to the hot water extraction machines used for carpets (see page 103).

General points to be remembered in connection with curtains:

– velvet and other pile fabrics should hang with the pile running downwards and they tend to hold dust and the smell of smoke.

- 15 to 30 cm, according to the type and weight of material, should be allowed for hem and turning on each curtain.
- when floor length, curtains should be 1.5 to 2.5 cm above floor level to prevent friction.
- the minimum width of any curtain should be at least 1½ times the width of the track, but for lightweight fabrics and certain curtain headings, twice or even more is necessary.
- curtain headings may be of various types, eg gathered or fix pleated, by the use of special tapes and the appropriate hooks or rings by which the curtain is attached to the rod or track (see Fig. 17.4).

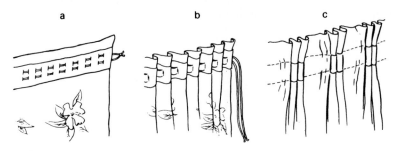

a　　　　　　　　b　　　　　　　　c

Figure 17.4　Use of special tapes for curtain headings

- the lining should be fixed to the top and the sides of the curtain, but the hem should not be attached.
- hems and sides of good quality curtains should always be hand sewn and never machined.
- heavy curtains may have weights or even a chain in the hems to improve the hang.
- it is necessary to use flame retardant fabrics in public rooms and for cubicle curtains in bed areas in hospitals and is advisable in all areas,
- draw cords or rods may be used to facilitate the pulling of curtains and to prevent the marking of the fabric (see Fig. 17.5).

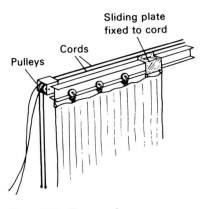

Figure 17.5　Draw cords

Non-drawing curtains

In some circumstances, for example where there is an opening into an alcove or with mock windows, venetian blinds etc, curtains may be needed for decoration and not made to draw. They are normally narrower than ordinary curtains and may be looped back at the side in various ways.

Pelmets, valances, swags

Pelmets and valances are decorative headings fixed over the top of the curtain to hide the suspension, to add decoration to the room, and in some cases to alter the apparent size of the window.

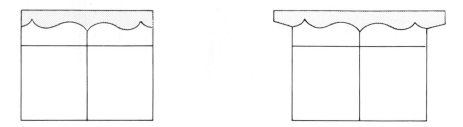

Figure 17.6   Effect of different length pelmets on the apparent size of a window

Pelmets are rigid, and may be of shaped pieces of wood or hardboard which can be painted to match the décor of the room, or they may be of padded plastic or stiffened fabric to match or contrast with the curtains. The fabric is mounted on stiff buckram and is often tailored to fit the window (Figs 17.6 and 17.7).

Figure 17.7   Curtains showing a pelmet

A valance is made of frilled or pleated material and hangs from a valance rail (see Fig. 17.8).

A swag is a draped finish to hide the curtain heading and is frequently completed with a tail (see Fig. 17.9).

Figure 17.8   Valances

Figure 17.9   Swag and tail

Net curtains

Net, sheer or glass curtains are made of translucent fabrics, frequently polyester net, which soften and diffuse the light as it passes through them. The curtains are used where windows can be overlooked, and are placed close to the window where they hold some of the dirt which would otherwise get on to the main curtains. They are held by rod or stretched plastic coated wire through the top hem; a drop rod is particularly useful for long net curtains, so that they can be lowered and changed easily without the use of a pair of steps.

With these lightweight fabrics the curtains should be two to three times the width of the window, and there should be sufficient weight at the bottom of the curtains to enable them to hang properly. This can be achieved by using treble hems, and with casement windows it is a good idea to have a rod or stretched wire through the bottom as well as the top hem, to prevent the curtains blowing outside, but the curtains must not impede the opening of the window.

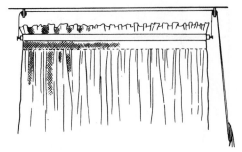

Figure 17.10   'Drop rod' for net curtains

Net curtains become soiled very easily and require frequent changing, so it is essential that there are at least two sets per window. Where there are different sized windows, the curtains should be marked.

There is considerable choice of materials for net curtains, white or pale coloured, patterned or plain, made of cotton or man-made fibres. It is wise to buy the best quality possible, as crispness and translucency are often affected by frequent washings in the poorer quality materials. Polyester net is to be recommended, as it withstands sunlight and retains its appearance after frequent washings. Only when the temperature of the washing water is above 40°C does heavy creasing occur which cannot be ironed out. Unless the curtains are stretched and held in position top and bottom, polyester net should be pressed after washing with a warm iron.

Other fabrics for sheer curtains include a net curtaining with a metallic backing. Aluminium particles are bonded onto a polyester net, so that 45–65% of the sun's radiation may be reflected.

Aluminium or steel wire mesh in a variety of colours will drape and can be used as curtains or partition material, as can metal chain.

Blinds

Venetian blinds are sometimes used in place of net curtains where windows can be overlooked and they can give protection from the sun to fabrics, paintings and objêts d'art. In some cases, ordinary curtains may be dispensed with, and the venetian blinds used with side drapes or strip framing, when there may be a considerable saving in materials.

Venetian blinds require constant care and cleaning, cut off a great deal of light if attached to the window head even when raised and guests complain of their noise. Vertical slat blinds are also available (see Fig. 17.11).

Many blinds of the roller type made of vinyl or similar coated material are available; they may be brightly coloured and patterned. When used on their own they do not give the well-finished appearance of a good curtained window but it is possible to use them with non-drawing curtains, thus saving material. If used to give privacy instead of net curtains they tend to obstruct the light, but they are much easier to keep clean than venetian blinds as they are easily dusted and may be taken down and wiped over when necessary.

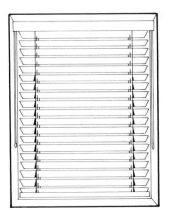
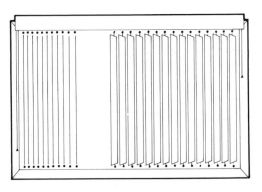

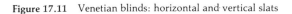

Figure 17.11 Venetian blinds: horizontal and vertical slats

Care and cleaning of blinds

1 Attend to badly hanging blinds.
2 Dust or wipe frequently.
3 Wash with warm water and synthetic detergent as often as required. This
 may be a contract job when they will normally be taken down and cleaned.

Austrian blinds are ruched fabric blinds. They are frequently used partially
raised and if fully raised they form a decorative pelmet. Austrian blinds have
gained in popularity in recent years, particularly in the public areas of hotels
etc.

Loose covers

Loose covers are detachable covers fitted over upholstered chairs, stools, etc.
They can give a clean, fresh appearance to a room, but the constant need to
straighten them and to keep them well maintained means that some establishments no longer use them.

Loose covers may be used to cover shabbiness, to protect the original
upholstery and to change the appearance of the room (eg winter and summer
coverings). The covers may reach almost to floor level with a pleated or
gathered 'skirt' or they may be tailored and fixed under the chair.

Fastenings may be hooks and eyes, zips or the 'touch and close' fasteners.
The last consists of two nylon strips, one with thousands of tiny hooks and
the other with thousands of tiny loops, and when pressed together the hooks
grip the loops to give a light and secure closure, yet can be easily peeled apart.
These fasteners can be washed or dry cleaned without damage.

Whereas almost any material can be used for curtains, closely woven
fabrics are to be preferred for loose covers, eg chintz, cretonne, etc. These
withstand abrasion, are less likely to snag, will not allow dust to filter
through to the upholstery beneath, and will hold their shape better than
loosely woven ones. A material which does not crease badly is an added
advantage.

Figure 17.12   Loose covers with a) a pleated skirt, b) a frilled skirt, c) a tailored cover with ties underneath

The choice of pattern, colour and texture is as for any other soft furnishing in that the material must be in keeping with the room and the rest of the furnishings. In some rooms material matching the curtains is used but normally this is better restricted to one or two pieces of furniture or it may become too dominant and, if patterned, the room appear overpatterned. Loose covers are sometimes considered a cheap substitute for re-upholstering but in reality they may cost nearly as much.

Loose covers may be laundered, but many are better dry cleaned as they are liable to shrink when washed and great difficulty will be experienced when putting them on again. There are nylon stretch covers which are easily laundered, do not shrink and because of their 'stretch' are easy to put on, but they may not be suitable for all shapes of chairs. All chairs and covers should be marked so that it can be seen to which chair a cover belongs.

To protect upholstered furniture from soiling in the most likely places, shields for arms and backs (antimacassars) are frequently used. These may be of matching material to the loose covers or of white or cream linen. They require to be washed frequently and need to be fixed firmly to the chair by special pins or the 'touch and close' fastening, or they present a very untidy appearance to the room.

Care and cleaning of loose covers

1 Shake and tidy frequently.
2 Brush or suction clean regularly.
3 Attend to repairs.
4 Have laundered or dry cleaned as required.

Cushions

Cushions may be used to increase the comfort of chairs and sofas and to bring colour, pattern and texture to the room. They may be shaped to fit the chair or sofa so forming the seat and/or back of the piece of furniture or they may be used loose as scatter cushions. Cushions may therefore be all shapes and sizes, filled with down, feather, kapok or foam plastic or rubber and covered

Figure 17.13 Fitted and scatter cushions

with a variety of materials, matching or contrasting with the material of the chair or sofa. The covers may be fastened in a similar way to loose covers.

Cushions require constant attention because they are often removed from their normal places, and feather ones all too easily become squashed and untidy looking and in need of repair.

Care and cleaning of cushions

1 Shake and tidy frequently.
2 Repair when necessary.
3 Brush or suction clean regularly.
4 Remove covers and wash or dry clean them as required.

Note Bedspreads and quilts are dealt with on pages 238–40.

18

BEDS AND BEDDING

When *guests* stay in an establishment they are naturally concerned with sleeping and the comfort of the beds is of great importance. The beds must not only be comfortable but must look inviting, and this will depend on the design, the materials from which they are made, and the neatly finished appearance of the beds in the room.

Standard sizes of single and double beds are:

$$100 \times 200 \text{ cm (3 ft } 3^3/_8 \text{ in} \times 6 \text{ ft } 6^3/_4 \text{ in)}$$
$$85 \times 190 \text{ cm (2 ft } 9^1/_2 \text{ in} \times 6 \text{ ft } 3 \text{ in)} \quad \} \text{ single}$$
$$150 \times 200 \text{ cm (4 ft } 11 \text{ in} \times 6 \text{ ft } 6^3/_4 \text{ in)}$$
$$135 \times 190 \text{ cm (4 ft } 5 \text{ in} \times 6 \text{ ft } 3 \text{ in)} \quad \} \text{ double}$$

There are also larger ones known as king size. As beds are individually made, it is possible for establishments to get a quotation for sizes other than the standard ones. It is usual in hotels for most rooms to have twin beds so that if necessary they may be let as singles, two people may share a room, or the beds may be made into a double bed by tying the legs or crossing the mattresses.

Construction of beds

A bed consists of a mattress supported by a base. The base may be of open coiled springs, wire mesh or laminated wood strips attached to a rectangular metal or wooden framework (see Fig. 18.1). The coiled springs may be padded and covered with ticking, or PVC for easier cleaning, to give an upholstered base; the wooden framework may either surround approximately half the height of the coiled springs giving a sprung edge, or the whole height when it gives a firm edge. This helps to prevent the springs sagging when the bed is sat on.

To prevent wear, and improve the appearance of an upholstered base, it may be surrounded by a fitted base cover or a valance may hang loosely round the edge; in some there are drawers or other storage space and these are useful in long-stay establishments.

When the base is not upholstered there should be an underlay of strong material, such as hessian or canvas, placed on top of the springs to protect the mattress from abrasion and for appearance the base may be surrounded by a

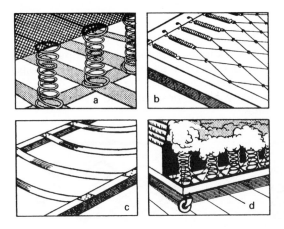

Figure 18.1 Bed bases: a) open coiled springs, b) stretched springs, c) laminated wood strips, d) upholstered

valance. Some bases have the edge of the framework raised so that the mattress is 'dropped' in and held in place. This type of base is useful in study bedrooms as it keeps a tidier appearance and also prevents damage to the edge of the bed when it is sat on. However, it does make bedmaking more difficult.

The height of the bed is not only important to the appearance of the room but also to the ease with which an elderly *guest* can get in or out of it and to the ease with which it can be made.

With the tendency for hotel rooms to be smaller than in the past the bed should not be so dominant that the room appears still smaller, so in many establishments divans are used instead of bedsteads. In the latter the legs with head and foot boards attached hold the base, whereas a divan is lower, with legs or castors fixed into the base, it may or may not have a separate headboard attached and is more easily converted into a settee by day. In hotels, divans usually have the headboards attached to the wall and the bed is pushed against it, while in hostels and similar establishments, where residents like to rearrange the furniture, the headboard is attached to the divan.

The headboard must be sufficiently high (30–45 cm above the top of the mattress) to protect the wall from soiling when *guests* sit up in bed. It may be made in different shapes and different materials but should be easy to keep clean and may incorporate a bedside table, light switches and maybe a telephone. The feet of any bed should be such that they will neither scratch a polished surface, nor damage a carpet by cutting or badly flattening it, nor move too readily when the *guest* is in bed. Suitable castors should therefore be used, such as small wheels which may lock, steel skids, etc.

In hospitals metal bedsteads are usual and they may be PVC coated. The headboard is normally adjustable to give a back rest for greater comfort and there may be mechanisms to raise the foot of the bed, or to enable the bed to be raised for making and clinical procedures, or to be lowered so that the patient may get in and out of it more easily.

Mattresses

In the past mattresses were stuffed and filled with hair, hair and wool or flock; but good hair is expensive, flock is cheap but becomes lumpy, and all stuffed mattresses are absorbent and liable to attack by moth and other pests. They require frequent turning when in use and remaking every few years and thus these mattresses have in many places been replaced by the interior sprung, latex or plastic foam ones.

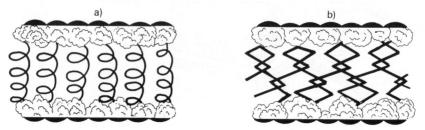

Figure 18.2　Sections through upholstered mattress showing: a) coiled springs, b) posture springing

Interior sprung mattresses are made of a continuous wire construction (posture springing) or of coiled springs of tempered wire very firmly held in position (see Fig. 18.2). There may be different grades of springs in different parts of the mattress, so giving more even support to the body when lying on them. The springs are well padded with layers of cotton waste, coiled hair, rubber or plastic foam, and the whole is then tightly covered with a strong ticking. Before the introduction of rubber or plastic foam into the padding, the cotton waste, wool and hair were held firmly in place by string pulled through at regular intervals and held by buttons, pieces of leather or tufts of wool on the top and bottom of the mattress. This tufting collects dust and now buttonless mattresses are usual.

Interior sprung mattresses should have a reinforced edge for added strength. They vary in depth (12–22 cm approx), quality and price, according to the number and gauge of springs, type of padding and quality of covering. To keep them in good condition they should be turned occasionally to even the wear. There are normally eyelet holes and handles on the sides of the mattress, the former to allow circulation of air and the latter to assist in turning. Interior sprung mattresses are heavy, absorbent and liable to attack by moth and other pests, but are comfortable and will last for many years.

Rubber mattresses are normally made from latex foam. The latex, which has previously been whisked with a chemical setting agent, is poured into heated moulds where it is shaped, set and vulcanized, without losing any of its tiny air cells. The mattress may be about 10 cm deep, and it normally has a right and wrong side due to the shape of the moulds and so should not be turned.

Deeper rubber mattresses have a specially developed foam base bonded on, but are still less deep and less heavy than interior sprung mattresses. As they are non-absorbent and return to their original shape after being lain on, they require no turning. They do not make dust or fluff, and are not liable to attack by moth and other pests.

Plastic mattresses are made from foam plastic, generally of the poly-

Figure 18.3 Divan with upholstered base and mattress

Figure 18.4 Bedstead with wire-mesh base

ethylene type. Foam plastic mattresses are non-absorbent, make no dust or fluff, and are not liable to attack by moth and other pests.

There is a danger that when foam rubber and some foam plastic mattresses catch fire they produce fumes which are toxic and although they may be treated this adds considerably to the price.

In hospitals both spring interior and foam rubber mattresses are used, the latter generally for patients who are up and about and the former for patients confined to bed. The mattresses have a Proban-treated 100% cotton cover which is inherently flame proof and satisfies BS 3120, as well as a polythene cover.

Cots and bed boards

In hotels, requests may be received for cots or bed boards. The housekeeper is notified by reception (if possible on the Arrivals and Departure list) and the house porter moves the articles to and from the rooms.

Cots are, of course, in use in hospitals also.

A *cot* consists of a mattress on a spring base and, to this, sides are attached to prevent the child from falling out. Appropriate sized sheets (often 'remakes' from the linen room), rubber sheets and blankets will be needed.

Bed boards are pieces of wood as wide as the bed and almost as long, which are placed between the bed base and the mattress.

Extra beds

Requests can also be received for extra beds, which are of two main types, ie zed-beds and sofa beds; but as foldaway beds are also a type of extra bed, they will be considered here too.

Zed-beds have a base of stretched springs which can be folded up into a narrow rectangular shape, enclosing a thin mattress (see Fig. 18.5). They can be easily moved and stored.

Extra beds which remain in the room without taking up space may be stowed away under beds (see Fig. 18.6) or may be put against a wall, giving the impression of a cupboard.

Sofa beds provide extra seating by day and a bed by night. A wooden slatted base combined with an interior sprung mattress is the most satisfactory type, as a wire mesh base tends to sag after prolonged use. Wooden

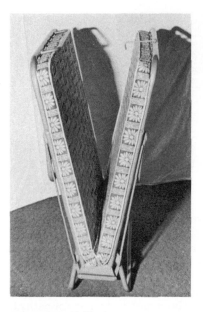

Figure 18.5   Zed-bed

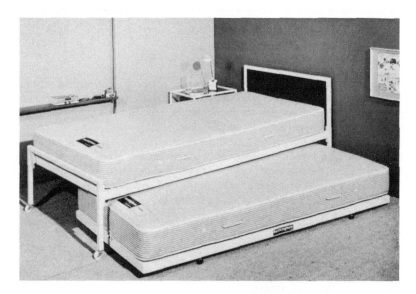

Figure 18.6   Foldaway bed

slats can be replaced if they break. A fitted sheet may remain on the mattress but blankets or duvet and pillows have to be stored in a cupboard.

Foldaway beds: when rooms are let more than once in 24 hours, for meetings by day and as a bedroom at night, a foldaway bed may mean that staff are not involved in heavy or time-consuming tasks in converting the room from one purpose to another

Figure 18.7   Foldaway bed

A foldaway bed folds into a wall, its underside giving the appearance of a wardrobe (see Fig. 18.7). The bed folds away fully made up and the fitting of the unit is extremely important.

Care and cleaning of beds

1 Check for loose headboards.
2 Check divan legs screwed in tightly.
3 Check mattress does not sag and buttons or other tufting are not missing.
4 Check for soiling and tears in ticking.
5 Turn interior sprung mattresses occasionally, both sides and top to bottom to ensure even wear.
6 Use underlays on bases of open spring type.
7 Supply waterproof sheet for young children and other necessary occasions.
8 Fit base covers or valances where bases not covered with PVC.
9 Dust or brush open wire springs occasionally and wipe with an oily rag.
10 Vacuum upholstered bases and mattresses and wipe plastic and rubber foam mattresses when required with a damp cloth.
11 Have valances and base covers laundered or dry cleaned when necessary.

Bedding

The term bedding applies to all articles on the bed and should strictly include bed linen also.

Pillows

Pillows consist of various fillings, covered with a strong closely woven material, ticking, which was formerly striped and is now more often white. The most usual size is 48 × 73 cm.

They may be filled with:

Down, which comes from the breast of the duck, is expensive and very comfortable, but is liable to attack by moth.

Small feathers, which are less expensive and less soft than down. They give a very satisfactory pillow but, like down, are attacked by moth. There are down/feather combinations.

Rubber or plastic foam (expanded polyurethane) } The comments given for the mattresses also apply to pillows (p. 230). However, many people find them too resilient for comfort.

Synthetic fibres, eg polyester, which are bulked to give a soft handle, are expensive, moth proof and extremely comfortable. They are less resilient than the foam ones and are used in hospitals.

Kapok, which comes from the cotton tree and not from the cotton plant, is soft at first but with use tends to become powdery and lose its softness and so is not a satisfactory filling; before the introduction of polyester filled, rubber or plastic foam pillows, kapok filled pillows were provided for *guests* with an allergy to feathers but are now not used. Kapok must not be confused with flock which, owing to its lumpiness, is not suitable for pillows.

Care and cleaning of pillows

1 Shake feather pillows daily.
2 Repair splits or tears in the ticking immediately.
3 Protect with an under pillowslip.
4 Have dry cleaned or laundered if necessary, with the exception of rubber and plastic foam which may be wiped clean.

Bolsters

Bolsters are elongated pillows which stretch the width of the bed. They form an underpillow, and as the head does not rest on them directly, they may be filled with a less resilient filling than pillows. They have gone out of fashion, and now a *guest* is normally given two pillows on the bed.

Bed linen

Bed linen should:

be comfortable,
have a good appearance,
be durable and withstand abrasion while on the bed and during laundering.

The strength of the material, its appearance and comfort depend on the type of fibre and weave chosen. Ideally the fibres should be:

stronger wet than dry,
smooth but not slippery,
absorbent,
cool,
able to withstand creasing.

The weave should be firm. A plain weave (alternate over and under of the warp and weft threads, see page 211) gives a firm material when the weave is close and balanced (ie little difference in the count of warp and weft). The count is the number of warp and weft ends per sq inch (6.45 cm²). A good quality cotton sheet will have a count of 140–180. The higher the count the better quality material and the less potential shrinkage there is likely to be. Shrinkage may occur over the first ten washes and may be as much as 5%.

Bed linen may be made from linen, cotton or synthetic fibres – see table below.

Characteristics of fibres used for bed linen

linen fibres	cotton fibres	synthetic fibres (nylon or polyester)
expensive, fresh, crisp feel, cool to touch, absorbent and comfortable, crease easily, retain good white colour.	less expensive, not so crisp, not so cool, absorbent and comfortable, crease less easily, keep fairly white.	prices vary, slippery (need to be fitted), limp feel, poor absorbency, clammy, less creasing, will discolour, strong wet as dry, withstand abrasion, soften with hot water and absorb grease, withstand alkalis and chlorine bleaches, drip dry, difficult to fold and store neatly, melt with heat of cigarettes.
stronger wet than dry, withstand abrasion well, withstand heat and friction of laundering, withstand alkalis and chlorine bleaches, require ironing.		

Besides linen and cotton sheeting, and materials made from synthetic fibres, bed linen may be made from flannelette or from a blend of linen/cotton fibres or polyester/cotton fibres.

Flannelette is a brushed cotton material and is cheaper and warmer than ordinary cotton sheeting but has not the crisp feel and appearance of linen or cotton. It is seldom used but some elderly guests request it in hotels and 'homes'. Flannelette sheets are sometimes used as underblankets.

Linen/cotton union is available but a material of *50/50 polyester/cotton* is now more frequently used. It has a better feel than 100% polyester and the potential shrinkage is less than for 100% cotton articles. It may be resin treated and then should not be subjected to temperatures above 80°C during laundering; it requires no ironing if folded warm from the tumble drier. This may be undertaken in an on-premises-laundry but not normally in a commercial one.

Coloured bed linen is used in some establishments but unless the same colour is used throughout it normally means extra work in the linen room sorting the articles.

Sheets should be long enough to give a good tuck in, and a good turn over at the top to protect the blankets and quilt from grease, newspaper print, the base of breakfast trays, etc. Normally for a single bed of 85 × 190 cm a single sheet should be 177 × 274 cm, for a double bed of 135 × 190 cm the sheet should be 238 × 274 cm, for a king-size bed of 182 × 200 cm the sheets need to be still larger and are frequently 274 × 297 cm.

For many years, sheets have had a wider hem at the top than at the bottom; this concentrates the wear in the same places and makes more work for the maids when making the beds and for these reasons many establishments buy sheets with equal hems.

The term 're-sheeting' normally includes putting out clean towels as well as making up the bed with clean linen.

Pillowslips will be made of the same material as the sheets. Frills and hemstitching are not recommended, and the housewife or flap type is the most usual as buttoned and taped slips need more attention regarding repairs. Even with the housewife style there is a tendency for the seam to become unstitched and hence the possibility of a torn flap, so establishments are now using longer slips without flaps, so that the pillow is hidden. Slips should fit easily over the pillow and are usually 50 × 76 cm.

Under pillowslips should be used to protect the ticking holding the filling of the pillow. They may be of cheaper cotton or be 'remakes' from redundant sheets; they are not changed as frequently as the top pillowslips, but laundered when they become soiled.

Duvet covers are large 'bags' of cotton, or more usually polyester/cotton fabric, into which the duvet or continental quilt is put. They are often patterned and the open part is held closed with press studs or touch and close fastener. The covers are approximately 135 × 200 cm for a single bed and 200 × 200 cm for a double.

Bath towels are usually of cotton in a turkish, ie terry, weave which has a looped pile on both sides. The pile should be close for greater absorbency and not too long or the threads will pull. The selvedge should be strong and the corners of the hems firmly stitched. The foundation cloth determines the durability of the towel. It should be strong, closely woven and when held to the light, little light should show through. A blend of polyester/cotton foundation cloth is sometimes used for strength, but it should be realized that the use of polyester will decrease the overall absorbency of the towel. Towels with fringed ends are rarely used as hems are stronger and stand up better to the frequent launderings.

Sometimes towels are coloured, but the colours should be fast so that they may be treated as white as far as the laundry is concerned. However, fast dyes do tend to discolour in time because of the alkaline detergent used in the laundering process.

There is a tremendous variation in the sizes of bath towels but 60 × 122 cm or 76 × 152 cm are frequently used, while larger sizes, eg 122 × 182 cm, are called bath sheets and normally only provided in private bathrooms in first-class hotels.

Face and *hand towels* may be of linen or cotton, and in the past were always of huckaback, which is a close, fancy weave and the best quality is very smooth and always made of linen. Now turkish towelling hand towels are being provided in the majority of hotels. The usual size of a face towel is

50 × 100 cm and normally one of these and one bath towel are provided for each guest. For use in cloakrooms, smaller hand towels may be provided and the sizes may be 30 × 45 cm or 25 × 35 cm; in many places, however, disposable paper towels and electric hand driers are taking their place.

Bathmats need to be very absorbent and are often made of turkish towelling or candlewick. These are laundered frequently and so are considered more hygienic than bathmats of cork or rubber. The size is 60 × 90 cm approximately. Disposable ones are available.

Blankets

Blankets provide warmth in bed and it is usual to provide one under-blanket (sometimes called bedpad) and two or three top blankets for each bed. The size of the blankets varies tremendously but they are generally a little shorter than sheets, eg 177 cm × 254 cm single, 228 cm × 254 cm double, as they do not require tucking in at the top. They are obtainable in a plain weave and 'fluffed up' by the use of a teasel or in an open weave, ie cellular.

White or pale coloured blankets are more often used in hotels, thus enabling the guests to judge the standard of cleanliness when they are sleeping in strange beds. The top or bottom edges of a blanket may be stitched with a white or coloured wool (blanket stitched), or have a coloured binding to match. This binding may be of a synthetic fibre and it gives a luxurious finish to the blanket. It is, however, expensive to replace when it frays.

Some reserve of blankets should be kept; these should be covered to keep them clean and if made of wool precautions taken against moth.

Blankets should have good insulation, ease of cleaning and non-flammability and may be made from wool, synthetic fibres or mixtures, or cotton.

Wool blankets are liable to attack by moth, are very warm and absorbent and are inherently flame retardant; they do not stand up to frequent laundering as they are liable to become harsh and felted so for hygienic reasons people sleep between sheets. Wool blankets get worn in time and the old ones are frequently used as underblankets; these are often folded and, if not long enough to cover the whole length of the bed, they should reach from the pillow to the foot of the bed.

Synthetic fibres, eg acrilan, nylon, etc, are light and warm and owing to their low moisture absorbency are very easily laundered without fear of felting. Acrilan blankets have been known to withstand 30 washings very satisfactorily and this could represent laundering twice a year for 15 years. Acrilan or nylon fibres are sometimes used in 'lace' blankets, eg Moonweave, when the blankets are extremely light in weight.

Mixtures made from wool/cotton, wool/rayon and wool/nylon are cheaper than wool, and more fluffy and attractive in appearance than cotton.

Cotton blankets are in a cellular form and are used in hospitals because of the frequent washing and boiling needed to prevent cross infection and are heated with a flame retardant finish to satisfy BS 5867 (Pt. 2).

Underblankets are not normally provided on hospital beds and the plastic mattress covers used lead to a build-up of static electricity and shocks may be felt when the metal bedstead is touched at the same time.

Blankets, especially under-blankets, require to be laundered or dry cleaned occasionally. All blankets can be washed, but woollen ones are liable to felt and shrink and so for preference should be dry cleaned; however this is too expensive for the majority of establishments. In order that blankets are maintained in a 'spotless' condition, a blanket book is sometimes kept to record their cleaning, as a blanket may have to go to be cleaned out of turn because of a spill or other accident.

Electric blankets (see p. 23) are provided in some hotels.

Care and cleaning of blankets

1 Take precautions against moth in storing woollen blankets.
2 Repair frayed ends.
3 Check for stains and dirty marks.
4 Shake occasionally.
5 Have laundered or dry cleaned when necessary.
6 Electric blankets should be returned to the manufacturer for cleaning and servicing periodically.

Eiderdowns and quilts

Eiderdowns are filled with down which, strictly speaking, should be from the eider duck but owing to the expense of down, they are often filled with curled feathers, man-made fibres or other materials, and should then be spoken of as quilts; the term *'quilt'* is now used to cover all types.

Quilts provide a warm, light bed covering but are too expensive, initially and in upkeep, for many establishments. In hotels with central heating it may be considered unnecessary to provide them in addition to blankets. They may be covered with fabrics which are slippery (these in the main are made from man-made fibres) and the quilts then tend to slip off the beds; in order to counteract this, the underside may be made from a less slippery material, eg cotton sateen or brushed nylon. Another method of securing them on the bed is to sew flaps of material on to either side of the quilt and to tuck these in, under the mattress. Quilts are generally placed under the bedspread and under the fold of the top sheet to keep them as clean as possible and to avoid tea and coffee stains which are difficult to remove from coloured fabrics. The sizes are approximately 84 cm × 127 cm and 144 cm × 127 cm.

Continental quilts or *duvets* filled with down or synthetic fibres are used in some establishments. They take the place of blankets and their warmth is rated in togs. Tog rating of 10.5 is average and the higher the rating the warmer the duvet is. They may be used with a bottom sheet only, when a clean cover for the quilt has to be provided for each new guest, and maids may find it more time consuming to renew covers than top sheets especially when there are a lot of 'one nighters'. To overcome these problems, in some hotels the duvet is used with a top sheet.

Duvets are not in general use by the Department of Health but may be found in some long stay hospitals, eg for the elderly. They have a polyester filling, a fire retardant primary covering which does not come off, a washable secondary covering treated to BS 3120, or of modacrylic or 100% spun polythene fibres, and a washable decorative outer covering. The recommended temperature of 71°C for 3 minutes for disinfection is not really suitable for polyester/cotton materials.

All materials having fire retardant finishes should be labelled so that the laundry can treat them as recommended.

Care and cleaning of quilts and duvets

1 Take precautions against moth in the storing of feather filled articles.
2 Attend to repairs.
3 Check for stains and dirty marks.
4 Have feather filled quilts professionally cleaned and others laundered.

Bedspreads

Bedspreads are used to cover the bed during the day and, where 'turning down' is still done, may be removed at night, folded and put away. The colour and pattern must suit the décor of the room, matching, contrasting or picking up a colour from the curtains or carpet. The bedspread is frequently sat on, has parcels put on it and may be folded nightly, so the material from which it is made should not crease or snag. Of the many materials suitable, some of the more commonly used are candlewick, cretonne, cotton satin, dralon velvet and tapestry and, in hospitals, heavy cotton fabrics with flame resistant finishes.

Figure 18.8 Fitted bedspread

Bedspreads may be of the fitted or throw-over type and in order that the corners of the latter do not hang too low at the foot of the bed they may be rounded.

The throw-over bedspread can be held in place under the pillow, and then carried over it to the back of the bed. Alternatively the pillow may be placed on top of the bedspread to which a reverse piece of material, about 114 cm in length has been attached and this is brought forward and tucked under the pillow.

Figure 18.9   Fitted bedspread brought forward over the pillows

Fitted bedspreads should be the exact size of the bed plus bedding and the sides may be quite straight, pleated or frilled to within an inch of the floor. The top of the bedspread can be finished as for the top of the throw-over type or if the pillows are removed and put elsewhere during the day, the cover can be made to lie flat on the bed.

In some cases where quilts are no longer used 'night spreads' of lightweight materials, or even a third sheet, cover the blankets and are tucked in with them to keep the blankets clean and to give a better appearance to the bed after the bedspread has been removed.

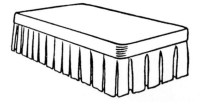

Figure 18.10   Fitted bedspread on divan with pillows removed

Care and cleaning of bedspreads

1 Attend to repairs.
2 Check for stains and dirty marks.
3 Shake occasionally.
4 Have laundered or dry cleaned when necessary.

Bedmaking

This constitutes a large proportion of the room maids' work in hotels and similar establishments, and a badly made bed can upset the *guest* and spoil the appearance of the room. An unmade bed gives a room an unkempt appearance, so beds are made as early as possible, and unless some thought is given to the method a great deal of time and energy can be wasted. The maid may walk round the bed many times unless she is taught to conserve her energy.

To make a bed

The following method has been found to be efficient:

strip the clothes from the bed on to a chair and leave to air;
turn mattress occasionally, unless made of latex or plastic foam;
working from the side replace underblanket;
put on bottom sheet, right side up and tuck in all round, making a mitre at
   all four corners. (*Note* To mitre a corner: tuck in along the foot or head
   of the mattress, lift flap of sheet from a point along the side about 30 cm
   from the corner, and tuck in remaining portion, drop flap and tuck in.
   See Fig. 18.11.)

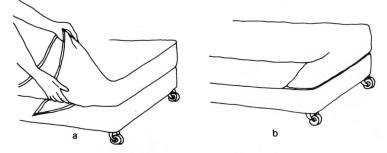

Figure 18.11   Mitred corner: a) in the process of being made, b) completed

put on top sheet, wrong side up, to reach just beyond the head of the
   mattress; (guest sleeps between the two right sides);
put on blankets separately, to reach just short of the top sheet;
put on quilt if used;
mitre one bottom corner and turn over sufficient of the sheet and blankets
   at the top to leave a space for the pillows, approximately 60 cm;
tuck in that side; repeat on other side;
replace the pillows with open ends away from the door;
put on bedspread.

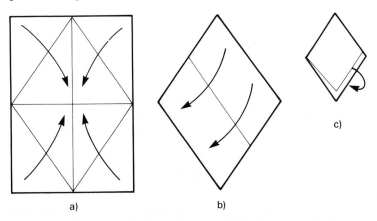

Figure 18.12   Stripping soiled linen from a hospital bed: a) corners to the centre, b) folded
in half, c) folded in half again

In some establishments duvets have replaced the top sheet, blankets and quilt and where *guests* stay several nights there may be some saving in time, but when the covers have to be changed frequently there may be little. They can give rather an untidy appearance to the bed.

In hospitals, and in particular in isolation and high risk areas to prevent the spread of infection when stripping the beds, the bedlinen is folded to the centre (see Fig. 18.12), then in half and half again.

There are other methods of bedmaking according to house custom and in some hotels the beds may be turned down when being made in the mornings and then covered with the bedspreads.

'Turning down' in the evening is an increasingly expensive process, owing to labour costs, and full night service is practised less and less, but it is still to be expected in luxury hotels (see Fig. 18.13).

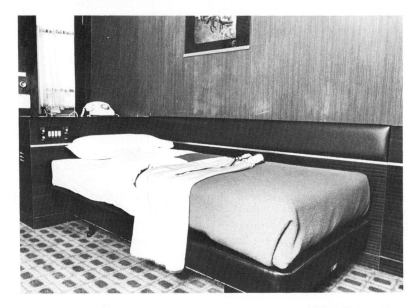

Figure 18.13   Bed turned down for the night

Turning down

This is one method which has been found efficient:

> remove and fold bedspread and put in a convenient place;
> untuck part of one side of the bedclothes and fold back the clothes to form a right-angled triangle;
> neaten the edges by folding under surplus bedding and tuck in at the side;
> place night attire on the bed, dressing gown on a chair and slippers at the foot of the chair.

Note   It is usual to turn down the side of the bed nearer the dressing table and if there are twin beds, the two inside edges.

Full night service in a luxury hotel will also include:

emptying any litter from ashtrays and wastepaper basket, and generally tidying the room;

wiping the bath and wash basin, paying particular attention to the toothglass;

folding the towels;

checking the lights;

adjusting the windows, drawing the curtains and later maybe putting in hot water bottles or switching on electric blankets.

19
FURNITURE

Choosing furniture

In any establishment furniture covers a wide variety of different items which will be in constant use and yet should retain their overall good appearance. People are seldom as careful of other people's property as they are of their own and the handling of furniture by large numbers of people results in harder use than if one person was using it all the time.

Practical design features may save space and money or simply make an area easier to clean and more desirable to use. Fashions change but at any time furniture needs to be:

practical in design, size etc.,
comfortable to use,
sturdy to withstand considerable wear and tear,
easy to clean and maintain,

and the price must be within the means of the establishment.

Whether the establishment is from the commercial or welfare field, whether it is luxury, medium price or of the lower price range, whether a holiday or city hotel and whether the *guest* is likely to be for a long or short stay, each piece of furniture must be fit for its purpose and meet the requirements of the *guests* and the *housekeeper*. The following points should be considered when choosing individual pieces:

type of guest expected and standard of accommodation,
guests' length of stay,
atmosphere to be achieved, eg modern, 'olde worlde', and degree of comfort,
shape and size of article in relation to the human body (science of ergonomics)
durability of article (it will be handled by a large number of people),
versatility and movability,
ease of cleaning, eg castors on heavier items, shelves instead of drawers, drawers with wipe-easy surfaces rather than lined, use of self-shine protective coatings,
standardization – items may be moved from room to room as required.

Where the type of *guest* staying in season may be different from the type staying out of season, the requirements regarding furniture may differ. In these cases dual purpose furniture may satisfy both types of *guests*. For example, in a university hall of residence a dressing table/writing table may satisfy the student in term time and the guest during the vacation or in a holiday hotel, the woman who on holiday has more time to 'make up' leisurely and the business person who has reports to write. More clothes storage space is required by the holiday maker than by the business person staying one or two nights, so a fitment which combines hanging and shelf space can provide sufficient space for the one and a neater piece of furniture than separate wardrobe and chest of drawers for the other.

Style, design and construction

The style of any piece of furniture must tone in with the rest, though it will not necessarily be of similar design. The whole should be in keeping with the style of the room. Chairs for dressing tables should be chosen with the particular dressing table/writing table in mind and any chair in the bedroom of a medium price hotel may not necessarily be suitable for a luxury hotel and vice versa. It may be an advantage to have a variety of easy chairs in a lounge, some with higher seats, or lower arms, or with wings, but each should be in keeping with its fellows, and none should look as though it had arrived there by accident.

Design and size are closely related to comfort, for inappropriate design or size may interfere with the proper function or the serviceability of an article. The width of the seat and the shape of the back of the chair are important to its comfort: the height of the table and the chair in relation to each other, the height and depth of the wardrobe and the length and width of the bed are other examples. Ergonomics is important when considering shape and size of pieces of furniture in relation to the body. Serviceability will also depend on design; shelves are probably more serviceable in the bedroom than drawers, and 'built-in' furniture can save space, labour, floor and wall coverings.

Flexibility and *movability* of furniture may be required in some places to enable the rooms to be put to different uses. Not only may it be necessary at certain times for guests' rooms in hotels to have extra beds or cots put in, to be let as twin, double or single but rooms can also be let for small functions, eg luncheons, exhibitions, or as syndicate rooms for conferences, when not only has the room to be set up in the morning but also returned to normal in the late afternoon or early evening.

Easily cleaned furniture, because of its design and the material from which it is made, is of importance in all establishments but especially so when there is the possibility of a quick turn round of rooms, ie a large number of departures in one day, as in hotels catering for tours and some holiday hotels, and in establishments in the medium and lower price range where maids may be expected to service more rooms.

In large establishments, the *quantity* of fittings required will probably enable them to be specially designed and ordered, but many places have to rely on ordinary domestic furniture which is not always suitable for the wear and tear to which it will be subjected. *Quality* will obviously determine

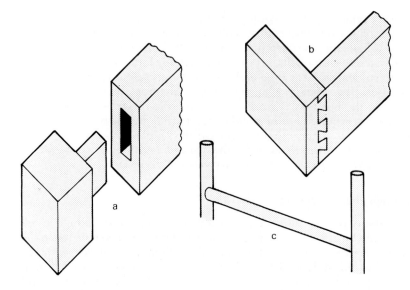

Figure 19.1   Methods of joining two pieces of wood: a) mortice and tenon, b) dovetail, c) dowels

durability, and the mark of the British Standards Institute and that of the Design Council can be of help when choosing furniture (see p. 91).

The method of *construction* and *materials* used will affect price, appearance and durability and the finished article should:

be free from rough, unfinished edges or surfaces,
be free from surplus adhesive,
have the correct type of joints which fit well (see Fig. 19.1),
stand firm on the floor and be rigid in use,
if a cupboard or wardrobe, be stable and balanced whether empty or full,
have drawers which run smoothly,
have doors which fit properly and have stays to prevent them opening too far,
have sliding doors which run smoothly,
have efficient locks, catches, hinges etc,
have handles conveniently placed, comfortable to hold and free from sharp edges,
have castors with no sharp edges (see Fig. 19.2).

Wooden furniture

Wood is the traditional and oldest material for furniture, and pieces dating as far back as 1500 still survive. Anything over a hundred years old is an antique, and beautiful antique pieces may be found in some establishments.

During the sixteenth and seventeenth centuries furniture was most frequently made of solid oak or walnut. About the beginning of the eighteenth century walnut veneers were being used; these were thin slices of wood about

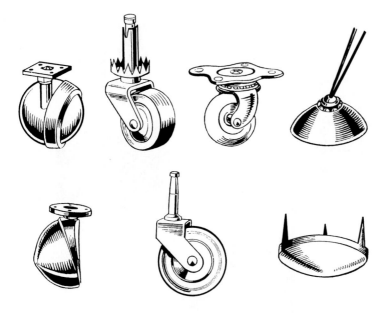

Figure 19.2 Variety of castors

1.5 mm thick (modern veneer may be 0.6–0.8 mm) glued over a carcass of the more common woods of that time eg oak, beech, elm, ash or pine.

During the eighteenth century mahogany became popular and the design of English furniture was influenced by such craftsmen as Chippendale, Hepplewhite, Sheraton and Robert Adam. By the beginning of the nineteenth century many other woods, including amboyna, satinwood and maple, were in use as well as the inlaying of metal, generally brass into wood.

The Industrial Revolution brought about the use of machines for the making of furniture and in 1877 William Morris tried to bring about a revival of craftsmanship; but machine-made furniture had come to stay and the Regency period gave way to the Victorian and a much larger and heavier type of furniture.

Today, there is a wide choice of woods eg beech, elm, oak, mahogany, African walnut and many others.

The more decorative woods are expensive and may be used as veneers, when an extremely thin layer is stuck on to a base of more plentiful, less expensive or more durable wood but which has not the same eye appeal. It is wrong to assume that veneered wood is of poor quality but the edge of the veneer may become easily damaged if sufficient care has not been taken during manufacture. Lipping, beading or framework can help in avoiding damage.

Solid wood is not always the most suitable material for a particular piece, or part of a piece, of furniture, and plywood or laminated wood often meets the requirements of modern furniture better.

Plywood is made by bonding together an odd number of thin slices or plies of wood, 1–2 mm thick, so that the grain of one ply is at right angles to that

Figure 19.3   Variety of chairs: a) early English, b) Chippendale, c) Regency, d) Victorian, e) modern

on either side of it and, since there is an odd number of plies, the grain of the two outside ones run in the same direction.

Plywood does not warp or twist to the same extent as solid wood, and is equally strong in both directions, whereas solid wood is strongest in the direction of the grain. Plywood is no cheap substitute for solid wood and is frequently used for table tops, eg 7 or 9 ply mahogany, where stability is required and for curved and shaped parts of furniture which can be preformed, so eliminating much nailing and glueing of parts and enabling stronger and cheaper pieces of furniture to be made. Laminated wood has strength in one direction and can be steam processed so that the arms and legs of chairs are made in one curved piece. Both solid wood and plywood may have decorative veneers on the surface.

Solid wood can be bent. Windsor chairs (see Fig. 19.4) and Bentwood chairs are examples of furniture made of bent pieces of solid wood. Beech is the wood most frequently used but birch, ash and poplar also bend easily.

Laminboard, battenboard and blockboard are more stable than plywood (as they consist of softwood strips sandwiched between two veneers); they require no framing and are often used for wardrobe doors. Chipboard materials are also used for furniture and these may have an integral melamine facing.

Figure 19.4   Windsor chairs

All wood should be guaranteed as properly seasoned and in the UK there should be about 12–14% moisture, or slightly less where there is central heating. As wood is extremely absorbent it requires some protective finish to prevent it absorbing moisture, grease and dirt and to make cleaning easier.

There are several protective finishes which may be given to complete the treatment of the wood. It is these finishes which determine the texture of the wood (ie whether it has a high gloss, dull gloss or matt appearance), its resistance to abrasion and the ease with which it can be cleaned. (See table below.)

Wood may be coated with paint when the natural appearance of the wood, ie grain, colour, texture, is lost. Paint provides a non-absorbent, easily cleaned finish in a wide range of colours, but it is easily scratched and has a poor resistance to heat. It is maintained by dusting, wiping with a damp cloth, or washing when necessary, avoiding the use of strong alkalis and coarse abrasives (see p. 194).

Wood finishes for furniture

Wax finish	*Oil finish*	*Nitrocellulose finish*
produced by rubbing in paste wax polish leaving a film of wax, applied to new wood or stained and filled surfaces, one of the most beautiful finishes, provides little protection, damaged by heat, water and alcohol, maintained by rubbing well when dusting and polishing periodically.	produced by rubbing new linseed oil and white spirit (4:1) into stained and filled wood, darkens the wood slightly, provides little protection – spillages are absorbed if not wiped up immediately, used on teak and afromosia, maintained by dusting and occasionally rubbing with linseed oil.	a varnish, ranges from matt to gloss, damaged by heat, water and alcohol, maintained by dusting and, if gloss finish, applying liquid, cream or spray-on polish occasionally.

French polish	*Acid catalysed finishes*	
produced by applying shellac dissolved in white spirit to the surface, high gloss obtained by rubbing the hardening shellac and repeated applications, scratched easily, damaged by heat, water and alcohol, maintained by dusting and periodically applying liquid, cream or spray-on polish, not often used now because of the cost of labour and the time involved.	produced by spraying or brushing a synthetic resin and a catalyst (the hardener) on to the specially stained and filled wood, the most permanent of all finishes, heat, water and alcohol resistant, maintained by dusting or wiping with a damp cloth. *Melamine* resin gives high gloss or finish resembling natural wood. *Polyurethane* has excellent water and chemical resistance. The satin finish has an exceptionally good feel. *Polyester* has a very high gloss and is the most difficult to repair.	

Care and cleaning of wooden furniture

1 Avoid scratching and knocking.

Figure 19.5 Wicker chair   **Figure 19.6** Example of metal furniture

2 Wipe all spills as soon as possible.
3 Treat stains as soon as possible (these are often produced as a result of spills not being wiped up quickly enough). (See pp. 143–4.)
4 Protect tops of dressing tables, coffee tables etc, with glass.
5 Examine for woodworm and treat accordingly (see pp. 146–7).
6 Clean regularly:
   a) Dust daily, rubbing well to improve appearance.
   b) If necessary, remove any stickiness or fingermarks with a damp cloth wrung out of warm water and synthetic detergent, or water and vinegar (one tablespoon to a litre of water).
   c) Periodically apply a suitable polish but not to a matt finish or it will lose its appearance and become glossy.

Wicker and cane furniture

Wicker and cane furniture is found in bedrooms and sun lounges of some places and is frequently painted. Unless well maintained it is liable to get out of shape, and pieces of wicker can protrude and catch on clothes.

Cane may be used for the seats and backs of chairs (see Fig. 19.5) and is occasionally seen on headboards of beds.

Both wickerwork and cane can become extremely dusty and shabby looking, if not well looked after.

Care and cleaning

1 Examine for broken and protruding pieces of wicker and cane and treat accordingly.
2 Clean regularly:
   a) Dust daily and use suction cleaner occasionally.
   b) *Wickerwork*: periodically wash, using a cloth or soft nail brush, warm

water and synthetic detergent, avoid using a great deal of water. Rinse and dry thoroughly. Polish with a liquid wax furniture polish.
Cane: periodically wash. Rinse with cold salt water and dry thoroughly.

Metal furniture

Metals in the form of iron and steel have been used for many years for bed-steads and the springs of furniture, but these and many others, owing to their strength and ease of shaping, are being used increasingly in modern furniture for the legs and frames of chairs (see Fig. 19.6) and for tables which have tops of such heavy materials as marble or ceramic tiles. Other metals used in furniture are light alloys of aluminium, chromium and brass.

To prevent corrosion many metals require a protective coating and this may be given by anodizing, electroplating, or the use of transparent lacquers, plastics, nylon, enamel or other chemical coatings and their appearance may then be maintained by daily dusting or wiping with a damp cloth and washing if necessary.

Some tables with metal legs, used in bars and lounges where drinks are served, have marble tops and it should be remembered that acid (eg lemon juice) eats into marble. To protect the marble, the surface may be treated with a catalyst lacquer.

Plastic

Plastics such as the catalyst finishes already mentioned, and synthetic adhesives which have considerable strength and the ability to bond together materials of differing compositions, are used extensively during the making of wooden and other furniture. Nylon and other plastic coatings are given to metal parts to protect them.

Laminated plastics are produced as veneers (see p. 158) under many trade names, eg Formica, and before being converted into furniture parts the veneers require sticking to plywood or similar supporting materials. They are used for table tops, dressing tables, bottoms of drawers (especially useful where cosmetics are likely to be spilt), wardrobes and similar pieces of furniture where durability and ease of cleaning are required. Because they have not the 'live' feel of wood, it is possible to compromise, by having the horizontal surfaces on which things may be spilt of laminated plastics and the vertical surfaces of wood. The appearance of laminated plastics may be maintained by dusting or wiping with a damp cloth. Abrasives should be avoided.

Reinforced plastics used for furniture are generally of the polyester glass fibre type, which can be moulded into seats and backs of chairs, and when used for the shell of upholstered chairs the chairs are much lighter in weight. As laminated plastics, these are durable and very easily cleaned.

Upholstered furniture

This type of furniture has become much lighter in weight as the methods of manufacture have improved. The older types had a framework across which

Figure 19.7 Hotel lounge area

webbing and hessian were stretched. Layers of suitable stuffing materials, such as tow, sisal, coconut fibre or hair, were then placed on the hessian and held in position by a piece of scrim or calico. So that there should be a reasonable amount of softness, there was next a layer of wadding, prior to the final covering which was stretched across the whole frame and its stuffing.

The next improvement was to increase the resilience of the upholstered piece of furniture by introducing coiled springs between the webbing and the

Figure 19.8 Chair with wooden tip to upholstered arm and loose seat

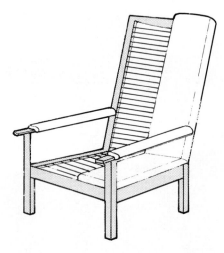

Figure 19.9   Chair with tension springs, loose seat and back

hessian. These added to the size of the article and many were large and heavy in appearance as well as in weight.

Tension springs are now often used instead of coiled springs. There is less padding and the seat is a loose piece of plastic or rubber foam cut to the size and shape of the chair and covered with the upholstery material. This provides a lighter chair and yet still fully upholstered. Show-wood tips prevent upholstered arms from getting quite so dirty (see Fig. 19.8).

Tension springs are also used in other easy chairs in which the wooden frame and arms are not upholstered; the springs are exposed and loose plastic foam cushions cut to size and shape form the back and seat of the chair. Because of the removable back and seat cushions the chairs are easier to clean and the loose backs and seats often have removable covers fitted with zip fasteners (see Fig. 19.9).

Upholstery materials

It is more difficult to judge the quality of upholstered furniture than of cabinet furniture, as so much of the material and workmanship is unseen. It is unlikely, however, that a good covering, well positioned, will hide faulty material or poor workmanship.

The covering will to a great extent determine the appearance, durability and cost of the piece of furniture and may be made from textiles, ie woven fabrics, hide or plastics. The covering is required to be:

resistant to abrasion, snagging, creasing, soil and fading,
non-flammable,
non-shading,
pest proof,
easily cleaned

and have stretch recovery.

Textiles

These requirements, in the case of textile coverings, will be met by the type of fibre, yarn and weave used in the production of the fabric. The performance of a fabric may be enhanced by the use of one or more of the fabric finishes which are now available (see pp. 213 and 216). Upholstery must comply with the Upholstered Furniture (Safety) Regulations 1980 regarding fire retardancy.

Smooth fabrics, eg brocades and damasks of cotton, rayon or synthetic fibres show soiling more but hold dust less than the rougher, textured fabrics of wool, wool/nylon and wool/Evlan mixtures. The latter have a warmer appearance, are less slippery and are less likely to show shine on clothes when sat on.

Cut and uncut pile fabrics, eg velvet, corduroy, or moquette, made of wool, cotton, rayon or synthetic fibres are hardwearing but hold the dust, and cut pile fabrics may show shading.

The use of synthetic fibres, eg nylon, dralon etc, either alone or in mixtures and blends increases the durability and ease of cleaning of many coverings.

Hide

Hide is durable and easily cleaned, but is expensive and must be kept supple to prevent cracking. It is inclined to be cold to the touch and to make clothes shiny when sat on, and for these reasons many pieces of hide furniture have loose fitted seats covered in some woven fabric.

Figure 19.10   Perspex chair with cushion

Plastic

There are many plastic materials available and some resemble hide very closely. They are more easily cleaned, equally hardwearing and less expensive than leather. They are normally vinyls and those with an expandable cotton backing are generally the most suitable.

Care and cleaning of upholstered furniture

Woven coverings
1 Watch for signs of wear and deal accordingly.
2 Remove stains as soon as possible (see p. 143).
3 Protect, if necessary, with arm covers, chairbacks or loose covers (see pp. 225–6).
4 Examine any woodwork for signs of woodworm and treat accordingly (see p. 146).
5 If possible, keep out of strong sunlight.
6 If not in use, take precautions against moth.
7 Clean regularly:
   (i) Remove ash and crumbs daily.
   (ii) Brush or use suction cleaner frequently, paying particular attention to the corners and sides of the seat.
   (iii) Remove stains as appropriate.
   (iv) Reverse seat cushions to even the wear.
   (v) Dust any showwood parts daily and polish periodically according to finish.
   (vi) Shampoo periodically as for carpets (see p. 189).
   (vii) Wash or dry clean any removable covers, when necessary.

Hide coverings
The care for hide or leather covered furniture is as for woven coverings above. Clean regularly:
   (i) Dust daily.
   (ii) Brush or use suction cleaner frequently.
   (iii) Periodically polish with good furniture cream to keep supple.
   (iv) If slightly soiled, wipe with a soft cloth wrung out of warm water and synthetic detergent. Rinse, dry thoroughly and polish.

Plastic coverings
   The care of plastic covered furniture is as for woven coverings above. Clean regularly:
   (i) Dust daily and polish showwood parts periodically.
   (ii) Brush or use suction cleaner if necessary.
   (iii) Wipe with cloth wrung out of warm water and synthetic detergent when necessary. Dry thoroughly.

Requirements for furniture

Chairs

Backs should be high enough to support the whole of the occupant's back. Seats should be long and wide enough to relax the thighs and knees. Depth of chair seat is related to the height of the chair, eg armchair 33–38 cm high seat, depth could be 60–70 cm, upright chair 42–45 cm high seat, depth could be 42–50 cm. An armchair should have a minimum width of 48 cm, a wing chair should have a minimum of 56 cm between wings. The gap between the back and the seat of an upright chair should be 20 cm high and backrest 20 cm, so

Figure 19.11   Chair with space between back and seat for easy cleaning

that the top of chair is 40 cm above seat, the gap enables easier cleaning (see Fig. 19.11).

Tables

The height for writing, above the chair seat, should be 28–33 cm; this allows for 18–20 cm knee clearance when the framing of the table is 10 cm. The height of the table from the floor is therefore 70–84 cm; coffee table 35–50 cm high.

Wardrobes

There should be: 60 cm minimum hanging space in a single wardrobe; 90 cm minimum hanging space in a double wardrobe; 56–60 cm depth to prevent rubbing of clothes when hanging. The height to accommodate full length dresses is 175 cm; a man's wardrobe may be 150 cm with 25 cm shelf for hats. The space above 200 cm is too high to be used conveniently. The hanging rail should be firmly fixed and not too close to the top, firm enough for heavy coats but not too thick for coat hanger hooks.

Fitted furniture

Furniture may be free standing, built-in, fitted or cantilevered. The first is self explanatory while built-in furniture often makes use of part of a wall, and generally cannot be removed without defacement, and fitted furniture is complete in itself but fixed in position. However, the terms 'built-in' and 'fitted' are used loosely to mean the same thing.

Built-in and fitted furniture save space, as well as wall and possibly floor coverings. They are normally simple in design and they enable the saving of time and labour during cleaning.

Cantilevered furniture is fixed on brackets to the wall and hence there are no legs to get in the way of cleaning.

Fashions in furniture change as in other things and in hotel bedrooms there is a move away from the fixed one-piece unit to individual pieces of traditional shapes which do not date and which perform their function well. They also allow rooms to be converted from one use to another more easily. Hoteliers can sell furniture when refurbishing and, provided it has been well looked after, a reasonable price can be expected after seven or eight years.

Leasing furniture

It is possible to lease furniture and the leasing contracts for hard furniture and upholstery in leather or expanded vinyl are generally written over a five-year lease period. For soft furnishings the period is shorter, generally three years. At the end of the period a new contract may be made for new furniture or the original articles may be kept at a reduced rental. A great advantage of leasing is that modernization is paid for with future profits, which the new furniture and furnishings will earn.

Once the particular pieces of furniture have been chosen and bought or leased, their arrangement in the room is of importance. The purpose of the room may be such that the furniture arrangement needs to be considered as a whole, eg in a bedroom, or there may need to be small groups of furniture, eg in a restaurant or lounge, but in all cases the arrangement should give a well balanced, inviting appearance to the room.

20

INTERIOR DECORATION (LIGHTING, HEATING, VENTILATION AND FLOWERS)

In the past, it was thought that the rooms in an hotel should, as far as possible, give the appearance of 'home from home', but it is now realized that guests, while still wanting to feel at home, expect something different in the way of decoration and that colours and designs suitable for the home often have a cold and unfriendly look in the impersonal atmosphere of an hotel. The trend for simpler architectural exteriors and simpler designs in furniture and furnishings has lent itself to the use of bolder and brighter colours. However there are vogues in colour and there has been some change towards softer tones and the wise use of colour is but one factor concerned in good decoration.

Colour

When choosing colours for any establishment it needs to be remembered that there are certain architectural and psychological aspects of colour; thus colours can substantially alter the apparent size or shape of a room, or add to its warmth, cheerfulness, peace and quiet,

eg reds, yellows, browns and the darker shades of most colours are warm and advancing colours and when used on an end wall may shorten the apparent length of the room or if on a ceiling may lessen its apparent height,

cool colours, pale greens and blues and lighter shades in general are receding colours and tend to make a small room look larger but should be avoided in rooms with northerly or easterly aspects because of their cold appearance.

While some people have a preference for one colour more than another, it is recognized that many colours have a similar effect on different people,

eg reds, oranges and yellows are warm and stimulating,
pastel shades are cooler and more restful,
green is cool and has a soothing, pleasing effect,
pale blue is fresh and cool but dark blue can be depressing if used in large areas,
purple has richness,

browns and other dark colours give the impression of comfort,
white can appear hygienic and cold.

So, in order that the occupants of a room should not be disturbed by the colours, the function of the room should not be forgotten when the colours are chosen. The entrance hall should look inviting, the lounge, suites and bedrooms restful, bathrooms clean but not cold (peach and pink may therefore be preferable colours to white); restaurants should have a relaxed atmosphere, while bars should be bright and cheerful.

Colours may unify an area but can be affected by:

the amount of light falling on them and so will appear different in areas of light and shade,
> eg curtains appear darker by day when light enters beside them than at night when the light falls on them;
> pile fabrics show shading according to the direction of the pile;
> an alcove, in shadow, may appear different from the rest of the wall;

the type of surface,
> eg a rough surface casts small shadows and so appears darker than a smooth, glossy one;

the surrounding colours,
> eg strong colours may distort other colours, they reflect on their surroundings,
> large areas of bright colours always appear brighter than small areas.

In most decorative schemes there will be a main, a contrasting and a neutral colour, and once the main colour is chosen according to the use of the area and the atmosphere required, a colour wheel may be used to help in selecting the final colour scheme. Neutrals, black, white, grey, cream, etc are not shown on the colour wheel, and can be used with any colour combination; see Fig. 20.1 below.

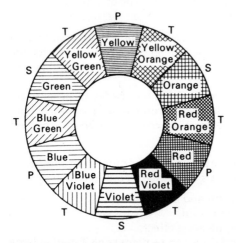

Figure 20.1   Colour wheel: P – primary colour, S – secondary colour (two primary colours mixed), T – tertiary colour (primary plus secondary colours mixed)

Thus, red
    yellow   } primary colours
    blue

    red and yellow = orange
    yellow and blue = green   } secondary colours
    blue and red   = violet

    red and orange = russet,
                burnt orange } tertiary colours
                coral etc

From the colour wheel the following colour schemes may be selected:

monochrome, using one colour in different shades and tints;

complementary, using contrast colours from directly opposite parts of the colour wheel, one in a bright tone for small areas and the other in greyed tones for larger areas. They should be used with a neutral colour;

split complementary, using a colour with the two on either side of its contrast; it may also include the direct contrast colour. The same rule applies as for the complementary colour scheme when it can give a very pleasing effect;

analogous, using colours which are related, ie are side by side on the colour wheel and the contrast colour of any one of the group can be used as an accent colour;

triad, using three primary, three secondary or three tertiary colours; one colour should be dominant and the other two in softened or greyed tones.

Pattern

Colour is frequently used in conjunction with pattern, and this adds interest to a decorative scheme, but the introduction of pattern is not without its problems, and needs careful consideration because:

it may help create the illusion of greater or smaller space;
too much pattern is disturbing to the eye and creates a 'busy' room;
large patterns can be overpowering in a small room;
small patterns may be lost in a large area;
if two patterns are thought necessary they should be different in character and one should be dominant;
the desired effect may be lost if:
    the pattern is seen only on a small sample,
    carpets are not seen lying flat on the floor,
    wallpapers are not seen hanging vertically,
    curtain materials are not seen hanging in folds.

Vertical stripes or any design which tends to move the eye upwards will make a room seem higher, and a narrow room look narrower. Similarly, horizontal lines make a low room look lower and wider. Some patterns have a three-dimensional effect giving an appearance of depth and may be useful in a small room.

Too much pattern is disturbing to the eye, so pattern should be used with restraint: thus a patterned carpet may be used with plain upholstery, patterned curtains with plain walls and vice versa. It is possible to introduce more than one pattern into a scheme, but they should be different in character – a striped and a floral pattern for example – and one pattern should always be dominant. The size of the pattern should be related to the size of the room or object and the patterned article should be seen in the position in which it is to be used.

Apart from any interest that patterns may add to the decoration, it should not be overlooked that patterned surfaces do not show marks and soiling as readily as plain ones.

Texture

In some instances texture takes the place of pattern and in decorative schemes where colour contrasts are not great, texture matters a great deal; thus different tones of gold or yellow may be used without any monotony when the upholstery is velvet, curtains silk and carpet wool.

Much more attention is paid to texture now than formerly and with the wide choice of materials available, variety in texture should not be difficult. In wall coverings alone, texture may vary from the cold, shiny, smooth surface of glass to the warmer, rougher surfaces of grass cloth, hessian and flock paper coverings.

For good decoration, therefore, the need is to try to choose colours, patterns and textures best suited for the particular room, and all three will be introduced into the room by the floor and wall coverings, furniture, furnishings and fittings. In any scheme it is usual to plan from the largest to the smallest areas; thus floors, walls and ceilings are considered first, doors, curtains and upholstery next, and the smaller areas and accents are considered last. In hotels a maintenance programme may be planned so that wall coverings are renewed every two to three years, soft furnishings every four to five years and a complete change of decoration will take place when the room is refurbished. Depending on the hotel this may be every eight to ten years.

Floorings

Floorings often outlast other furnishings, so many decorative schemes have to be planned to fit in with the existing floor. The tendency is, where possible, to have fitted carpets throughout the 'house' as these:

provide only one floor surface to be cleaned,
give a warmer appearance,
seem to add space,
make for easier furniture arrangement.

A carpet square or a number of small rugs tend to:

break up a floor area,
reduce room space.

Small carpets or rugs particularly tend to:

separate related furniture groups into distinct units,
bring colour, pattern and texture to an otherwise plain floor.

A patterned carpet should have a design in keeping with the size, style, function and atmosphere of the room and is normally chosen:

for large rooms and rooms with plain walls and/or upholstery,
where soiling and staining are to be expected,
where the floor has to give a good appearance at all times.

A plain carpet is suitable for:

small rooms,
rooms with patterned walls and/or upholstery,
giving an appearance of spaciousness.

The colour and tone of the carpet should be such that it unites the whole room.

Wall coverings

In many decorative schemes wall coverings are not dominant; they more often than not form a background for the other items. However, this does not mean that they are without colour, pattern or texture, which must, as for everything else, be suitable for the size, style and function of the room. There is probably a wider choice of wall coverings than any other item in the room, when one takes into consideration the different materials, colours, patterns and textures used during their manufacture (see Chapter 15).

Mirrors

Framed, unframed or in the form of tiles, mirrors are often found on walls. Due to reflection from their smooth, shiny surface they:

may be a foil for less shiny surfaces,
make a room appear larger,
increase the light,
add to the appearance of a vase of flowers or similar object.

Decoration can be increased by designs being etched on the glass and tinted mirrors, which give a warmer reflection than the more ordinary silvered ones, are also available. Careful consideration should be given to the positioning of mirrors on landings, because, although they can be useful in increasing light and the appearance of space, they can be a source of danger to short-sighted and absent minded people.

Curtains

When drawn at night, if the windows are large, curtains may become one wall of the room and when of a similar colour to the other walls they can produce a very restful effect. The only contrast, in this case, is the texture and the folds of the fabric and, as the continuation of colour increases the sense of space, this treatment is particularly suited to small rooms.

Alternatively, the curtains may be chosen to be in contrast to the walls, when they become dominant features in the decorative scheme. Strong pattern, colour and texture add importance to the window and can detract from the size of the room and it must be remembered, when the windows are large and it is decided to make the curtains the main feature of a room, that the rest of the scheme must be restrained because the area of pattern or colour is large and not broken up in the same way as that of walls and floors.

The length of the curtain track, the length of the curtain and, where used, the type and size of the pelmet or valance may affect the proportions of the window and consequently the room, eg the window appears wider when the curtain track and curtain extend beyond the window frame, and full length curtains may take from the width of the window and make the room appear higher (see Fig. 20.2 below).

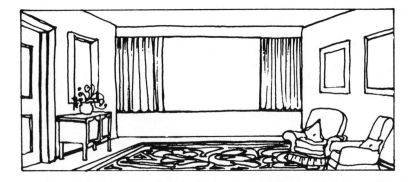

Figure 20.2   Effect of different curtain arrangements

Pelmets are generally one-eighth to one-sixth the height of the curtains and the deeper they are the less high the windows appear. Their use as a decorative feature of the room will depend on the material from which they are made, their colour (they may match or contrast with the curtains), and their shape. When there are no pelmets or valances the heading of the curtain is of much greater importance (see Chapter 17).

Furniture and other furnishings

While floor and wall coverings are important features in the décor, furniture and other furnishings must not be forgotten. For a room to appear comfortable it must contain furniture that:

is functional
does not lose the sense of space in a small room
blends in with the rest of the decoration.

Wood used in furniture has a warm appearance and is normally regarded as a neutral, but certain pieces of furniture, owing to the colour or grain of the wood, may be made to stand out from the rest; further contrasts in colour and texture may be introduced by incorporating metals, glass, marble, plastics and fabrics as part of the furniture. Timber-look laminates are used extensively, especially for bedroom furniture, and are better in appearance than before. They are easy to clean.

Where the furniture is upholstered or loose covers (including bedspreads) are used, the same care should be taken as for curtains when choosing the colour, pattern and texture of the fabric. Sometimes the same fabric is used for loose covers as for curtains and the best effect is probably obtained when only one or two pieces are covered in the same fabric. Textiles are very important to decoration as they provide a large part of the visual appeal and comfort in a room.

A bed normally has a head rest or head board which in the case of a divan may sometimes be fixed to the wall instead of the bed itself. From a decorative point of view it should be of a suitable material, shape and colour to fit in with the rest of the furnishings, while from a practical point of view it should withstand frequent cleaning and be high enough to protect the wall surface when the *guest* sits up in bed. The height of the bed is important to its appearance in the room.

A scheme for any room may be softened or accents of colour can be introduced by the use of pictures, cushions, lamp shades, flowers and other accessories. Even such items as waste paper baskets and ashtrays should be considered with the scheme as a whole.

The size and colouring of any pictures chosen will depend on the wall space available, and the general décor of the individual room. The choice of subject, however, is not easy as personal prejudices enter into it. Landscapes and floral paintings probably appeal to the majority of people. Pictures may be sprayed with a plastic which does not alter the colours at all but renders the pictures washable and enables the glass to be dispensed with; this may be an advantage as glass has a tendency to cause reflections and can get broken. The frame should set off the picture and be in keeping with the style of the room. Pictures are frequently hung too high; they should normally be about eye level and there should be no cord or wire showing. It is possible to hire pictures and in most hire schemes the pictures are changed about twice a year.

There are some rooms which are used mainly in artificial light and disappointing results have been obtained when the décor has been planned in daylight. The majority of rooms are, of course, seen in both artificial and daylight. It must be remembered that parts of the room in shadow during the day, eg curtains, may not be so at night and for good decoration at all times much will depend on the type and positioning of the lights.

Lighting

Lighting plays a key role in creating the right atmosphere within an area. It should be decorative as well as functional; it should contribute to the character and atmosphere of a room and be adequate for general and particular purposes, without causing glare or appearing flat and dull. To achieve this in any given room the direction and quantity of lighting have to be chosen for the right effect and function. There normally has to be a balance between direct and indirect or direct and diffused lighting systems.

In the case of direct lighting, the fittings throw the light onto surfaces below, generally producing over-bright areas with hard shadows, resulting in glare and highlights on polished and other smooth surfaces.

For indirect lighting, the fittings are concealed and the light is thrown on to the ceiling and walls, from where it is reflected into the room. No glare or hard shadows are produced but the lighting tends to have a 'flat' appearance and is very much less economical in use.

When the fittings are completely enclosed or concealed as with some globes, ceiling panels or laylights, the light is diffused as it passes through the glass or plastic fitment. Diffused lighting is also glare-free and produces a 'flat' appearance. It is possible to have some light passing through a diffusing bowl and some reflected from the ceiling; this is semi-indirect light (see Fig. 20.3 below).

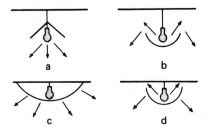

Figure 20.3   (a) direct lighting   (b) indirect lighting
(c) diffused lighting   (d) semi-indirect lighting

The amount of light required in an area depends on the function of the area and on the predominant colour used in the furnishings or floor coverings (dark colours absorb light and light colours reflect it) and in areas where strong, clear light is required the overall decorative lighting may be supplemented by purely functional lighting.

In any establishment management is not only concerned with the aesthetic aspect of lighting but also with low energy costs and ease of maintenance and the types of electric lamps (bulbs) need consideration. Electric lamps may be filament or fluorescent – see table below.

Characteristics of filament and fluorescent lamps

Filament lamps	Fluorescent lamps
vacuum-filled or filled with inert gas,	vacuum-coated with fluorescent powder of various colours,
fine filament inside heated to incandescence,	ultraviolet radiations fall on the powders to give light,
coiled coils have higher operating temperature and more glare than same wattage single coil,	wide choice of colours available,
average life 1000 hrs, long life 2000 hrs,	some tubes have reflectors inside,
life reduced if not suspended from cap, by vibration, wrong voltage or variations in mains pressure,	hot cathode lamps require low starting voltage, so more convenient for general use,
bayonet fitting to 150 watts, screw cap above 150,	cold cathode lamps start instantaneously, last longer but are more difficult to instal as require high voltage,
low in cost, different sizes suitable for many types of fittings,	average life of hot cathode lamp 5000 hrs, life of cold, 15 000 hrs,
are tubular-shaped or architectural lamps 30–56 cm long,	variation in length of tube 15 to 244 cm long, large areas can be lit from a single point,
give direct (clear) or diffused (pearl) light of varying intensities,	initial installation cost higher than filament,
can produce sparkle and highlights,	running costs lower,
give warm light with little distortion of colour,	give 3 times as much light as filament lamps of equal wattage,
generate a considerable amount of heat,	give diffused lighting with flat appearance, textural interest and highlights can be provided by supplementing with filament lamps,
will soil and stain walls and ceilings,	
are easily replaced,	
particularly suitable for pendant lights, wall brackets, table and floor standards, 'spot' lights,	operating temperature is one-fifth lower than filament lamps so not the same degree of soiling,
used with many different types of fittings and shades,	give good overall general lighting with no shadows,
poor efficiency compared with fluorescent lamps.	particularly suitable for cornice lights and obscure corners,
	not same demand for shades as for filament lamps and used most frequently with plain plastic or glass covering the whole length of the tube.

Low-energy bulbs are available which, although more expensive to buy, do offer considerable savings in energy costs. Energy-efficient compact fluorescent lamps are steadily replacing ordinary light bulbs. They consist of a small fluorescent tube bent into a convenient shape to fit into the conventional lamp holder and its design permits an elegant range of fittings.

Low-voltage down lights are another means of saving energy and are particularly suited to public areas. They give positive pools of light, enabling certain features of design to be accented and they have a low heat emission.

Reflector type incandescent (filament) lamps are used extensively for feature and decorative lighting. These are usually 100 or 150 watt rating and have silvered or aluminized inner reflectors. Some of these have a pressed glass front and can be used equally well for indoor or outdoor use, provided the lamp holder itself is weather-proof. In tungsten iodine lamps used for floodlighting, there is no blackening of the lamp envelope and hence it can be reduced in size.

Fittings and shades

Many electric lamps, particularly filament ones, need to be concealed as much as possible, when the shades and fittings become of great importance to the efficiency of the lighting system and the decoration of the place in which they are to be used.

Filament lamps may be used in pendant fittings, wall brackets, table or standard fittings and each one, with its appropriate shade, may be chosen from a very large number of shapes, sizes and materials. Standard lamps should have a firm base and the shade should be in the correct proportion to the height of the lamp. Where there is any doubt as to size the shade should err on the large side rather than the small. Lengths of flex should be avoided across the floor, over the furniture and under carpets; they are unsightly and may lead to accidents by people tripping over them or by their becoming worn and the consequent danger of fire.

Shades may be of glass, plastic, parchment, fabric or even metal, which is generally cellulosed and sometimes perforated. Some materials tend to fade, discolour or become ruined by the heat more than others, but all, in time, become dirty and not only is the dirt unsightly but it takes from the efficiency of the light. There are instances where the colour of the shade required for the décor casts an unsatisfactory light and the shade should then be lined. The shade should be pleasing to look at whether lit or unlit. Shades may be dusted or suction cleaned and many may be immersed in warm water and synthetic detergent when special thought should first be given to any trimmings on the shade.

To satisfy the *guest* and management light fittings should be:

well positioned,
emit light in direction and quantity required,
pleasant to look at whether lit or unlit,
well made, durable, mechanically sound and electrically safe,
not prone to overheating,
such that lamps can be replaced easily,
easily cleaned.

The amount of light emitted is measured in lumens. The minimum lighting requirements are as follows:

Restaurant	7 lumens/929 cm^2
Bathroom	10 lumens/929 cm^2
Stairs	10 lumens/929 cm^2
General offices	15–20 lumens/929 cm^2

A wall fitting

A pendant fitting

Concealed lighting

Figure 20.4 Different types of lighting

A table lamp

Casual reading	15 lumens/929 cm²
Sustained reading	30 lumens/929 cm²

Uses of lighting

Lighting may be used in interior decoration to:

reveal features of construction,
conceal space by areas in shadow,
create impression of space,

create suitable atmosphere in room or area,
act as a focal point,
accentuate colour and texture,
highlight pictures, statues, floral decorations etc,
introduce accents of colour by fittings and shades.

Strong shadows may be avoided by using multi-light fittings rather than one single light of equal wattage. Lighted walls:

give a background to furnishings,
give added perspective and feeling of space,
can dramatize colours and textures.

Cornice lighting gives downward light, lights the wall and emphasizes ceiling height.

The *entrance hall* to any establishment should look inviting and the lighting should be in keeping with the character and atmosphere of the place. During the daytime an entrance can appear dull and dim when the *guest* comes in from outside.

In a large area a chandelier or other pendant type fitting may give general lighting, or there may be overall lighting of the ceiling by means of cornice lights when the light will be reflected from the ceiling which should therefore be light coloured. General lighting of the area may also be provided by wall brackets or pelmet type fittings. (See Fig. 20.4.)

If height permits, a false or suspended ceiling may be constructed of various materials, and it may be possible to have glass panels or recessed downlights in the ceiling (see Fig. 20.5 below). An advantage of a false ceiling is that the lamp fittings are concealed from view.

Figure 20.5　Illuminated ceiling

Mirrors on the ceiling reflect light and give the impression of greater height and may provide an interesting reflection of the light fittings.

In the entrance hall there should be areas of brighter light to attract *guests'* attention to such places as the reception desk and to enable them to see clearly to sign the register (see Fig. 20.6).

Figure 20.6   Well-lit reception desk

The atmosphere in the *lounge* should be one of comfort and restfulness and as much consideration should be given to the lighting as to the furnishings, bearing in mind that not all parts of the room require the same degree of illumination.

Recessed downlights and wall lights provide some degree of general illumination without the appearance of brightness and may be used as local lights when necessary (see Fig. 20.7). In the case of wall lights the light is, however, limited to the perimeter of the room unless there are pillars to which brackets may also be attached. Portable fittings (floor and table standards) are useful and providing there are sufficient sockets available these may be

Figure 20.7   Recessed lights in a cloakroom with textured ceiling

placed in different positions in the room. (Floor sockets, covered when not in use, prevent the trailing of flexes.)

There are many types of places in which food is eaten and the atmosphere will vary accordingly. In places where there is a cafeteria service or a quick turnover it is usual for there to be a high degree of illumination, especially at the counters and tables, and in these circumstances fluorescent lamps, pendant fittings or panel lights may be used.

In the *restaurant* of an hotel, subdued lighting is more usual, especially at night when table lamps or even candles may be thought necessary. Higher general lighting is normally used for banquets and luncheons but, even so, fluorescent lighting is not often the only source of light as it provides no high-lights, and due consideration should be given to its effect on the colour of food.

Subdued lighting may be required in the *corridors* but gloom should be avoided, and *guests* should be able to see the room numbers clearly. The space between the light fittings along a corridor should not be greater than 1½ times the distance they are above the floor. Stairs should be well lit to prevent accidents, and lights can be set into the stair itself or along the wall just below the handrail (see Fig. 20.8). If the lights are overhead, the fittings should be placed at each end of each flight of stairs.

Figure 20.8   Staircase lit from below the handrail

For safety reasons, corridor and staircase lights should be left on during the night and there should be secondary or emergency lighting operated from an entirely independent emergency supply for corridors, staircases, banqueting rooms and for exit signs.

Bedrooms do not necessarily require general lighting but there should be adequate light in the different parts of the room. There should be a bedside light for each bed, as well as a dressing table light. Bedside lights may be table standards or mounted on the wall, but in each case they should be sufficiently high to light the book being read and not the pillow, top or side of the bed (see Fig. 20.9). It is a good idea to have two-way switches to be worked from the

Figure 20.9 Good bedside lighting by means of: a) overhead filament light, b) bedside lamp, c) overhead fluorescent light

door and the bed. In the case of double and twin beds it is kinder to have two lights for reading in bed, each with its own switch, rather than one light only. Dressing table lamps should light the face and not the mirror; two lamps are very suitable or one long fluorescent lamp above the mirror. For a study bedroom or studio room there should be a good light which is adjustable for writing and reading, and for these purposes an 'Anglepoise' type of light is suitable but not so decorative.

In *bathrooms* there should be vapourproof fittings, and the position of the light by the mirror should be such that the face is lit adequately. The switches should be either outside the room (not for public bathrooms) or of the cord type.

Heating and ventilation

The comfort of the human body is dependent on its being surrounded by moving air of a suitable temperature, humidity and composition, and it is important that these conditions should be produced with the minimum amount of cost and inconvenience and with some consideration for the design and decoration of the place.

Individuals vary as to their idea of 'suitable' conditions but it is generally accepted that there should be:

a temperature of between 15°–20°C,
a relative humidity of between 40–60%,
not less than 2,800 cm^3 of fresh air per person per hour.

Recommended standards of warmth are:

lounges 20°–21°C,
bedrooms 13°–16°C,
general offices 20°C,
factories, depending on work, 13°–18°C,
general spaces, entrances, stairs etc 16°C,
lavatories 18°C,
hospital wards 19°C.

In residential accommodation for hospital staff the Department of Health recommends a background temperature of 13°C and that the difference be made up with secondary heating. This is thought to be more economical as rooms are often not in use.

The occupants of a room contribute continually to the temperature and humidity of the room and where there is a large number of people more changes of air per hour may be necessary if the occupants are to feel no discomfort. The Department of Health recommends that for nurses' accommodation there should be 1½ changes per hour in a bedroom and 2 changes per hour in a common room.

It is usual to expect the number of air changes per hour to be, in:

kitchens 20–60
bathrooms 6
cloakrooms 2

In most places, bedrooms and lounges etc are ventilated naturally by the use of the windows, but for such places as bathrooms, cloakrooms, restaurants and kitchens some mechanical means may be necessary to introduce fresh air and/or extract stale air. In the case of internal bathrooms in London extract ventilation is essential to conform with by-laws, but there need be no fresh air inlets. Great care should be taken in the siting of the inlets and outlets to prevent draughts, the feeling of stagnation and unsightliness.

In most establishments, there is some form of central heating and the circulating medium may be hot water, steam or warm air produced by the combustion of solid fuel, oil or coal gas, or by the use of electricity. In many systems radiators are required. The position and type of radiators or grilles need careful consideration from the point of view of appearance, ease of cleaning and efficiency of the system. The trend is for the radiator to be enclosed behind a grille; thus difficult and constant cleaning is avoided and the arrangement of furniture is easier. Grilles for warm air should be placed so that they are as inconspicuous as possible.

In modern establishments full air conditioning plants are installed when filtered air, at controlled degrees of temperature and relative humidity, enters the individual rooms. Room inlets, mostly in the form of grilles, are generally made of metal but plaster or wooden ones are available when required to be used from a decorative point of view. Extract gratings remove the air from the room when up to 66 per cent may be recirculated after being mixed with at least 34 per cent fresh air and warmed or cooled, dried or moistened as required. Most air conditioning systems work best when the windows in the building are closed, and so eliminating noise.

In buildings not fully centrally heated or where there is no central heating, other forms of local heating appliances, such as radiant and convector gas or electric fires, will be provided in particular places, for the comfort of the *guests* and staff. These should be chosen for their appearance as well as their efficiency. The efficiency of a radiant electric fire is dependent on the shininess of the reflector and so this must be kept dust free and polished. Heating units incorporating air cleaning facilities are available and may be used in bedrooms etc where air conditioning is not yet installed. All open fires, be they solid fuel, gas or electric, must be adequately guarded.

In public rooms and suites, coal fires aid ventilation, are cheerful and give a pleasing focal point, providing the fireplace is in keeping with the rest of the décor. Coal fires heat rooms unequally and are inadequate for the heating of large rooms without some form of supplementary heating. This, and the fact that they involve a great deal of work, account for their decreasing use; in most instances now their main purpose is as a focal point and not for heating.

Energy saving

Space heating accounts for 45% of total energy spent in an establishment, half of which is lost by transmission through walls, windows, roof and floor of the building.

Energy may be saved by:

use of thick curtains, fitted carpets, close fitting doors and windows, entrance lobbies, edge sealing of windows and reflective material behind radiators,

use of timing devices and thermostats, including thermostatic valves on individual radiators,

controlling the temperature of domestic hot water, 40–45° for bathrooms, 60°C for kitchens,

encouraging the use of showers,

use of spray taps in cloakrooms,

isolating areas which may not be in use (using computer controlled systems),

good maintenance of taps, heating and dish washing equipment,

ensuring efficiency of the heating source and system, eg regular maintenance, no obstruction of radiators and heating ducts, lagging of pipes etc.

Flowers

In some establishments flowers are used extensively and there may be a large arrangement of flowers in the entrance hall or foyer, flowers in the lounge, restaurant and the suites; in other places there may be only one or two arrangements, but as a rule *guests* are appreciative of the time and trouble spent on the arrangements, and of the pleasing atmosphere they provide.

The arrangements should match the standard of the establishment as well as complement its style, décor and furniture. Flowers need to last and this limits their choice; they may often be spoilt by guests disposing of unwanted drinks (eg in bars) and the lack of light in many areas (which is more of a problem for pot plants than cut flowers).

The extent to which flowers are used in any place will vary: first, with the degree of luxury, when the arranging of personal flowers, such as button-holes, sprays and the ordering and despatching of flowers for special occasions, may be a part of the service offered to the guests in first-class hotels; secondly, with the number of special functions; and, thirdly, with the house policy.

As the *housekeeper* is responsible for the appearance of most parts of the house, she is naturally concerned with the flowers, but the extent of her responsibility for them will vary from one place to another. The *housekeeper* may be solely responsible for the flowers, doing the work herself or delegating it to an assistant; or a part-time florist may have a regular appointment to arrange large displays etc, and the *housekeeper* undertakes the everyday maintenance of the displays, bedroom flowers, etc.

Where a first-class hotel has a full-time florist or a florist's shop on the premises, the flowers are not the responsibility of the housekeeper, but even so when guests bring in or have flowers sent to them it is usual for the housekeeper to provide vases and, when requested, to arrange the flowers.

Floral arrangements may be provided on contract, when the arrangements are brought in and taken away at agreed times and little or no work is done on the premises.

An enterprising florist will use unconventional containers, and will employ things other than flowers to vary the decoration, such as pieces of wood, bark, stones, fruit and sometimes even vegetables.

Unless ready prepared arrangements are brought in by a contractor there needs to be a flower room with sink and running water, containers in which the flowers may soak, a good assortment of vases of different shapes and sizes, scissors, wire netting (3.8 cm mesh), plastic coated to prevent rusting, etc. The smaller items, together with dried grasses and all the 'bits and pieces' associated with Christmas decorations, are best kept in a drawer or cupboard. The photographs in Fig. 20.10 below illustrate the use of varied containers and flower arrangements.

Figure 20.10   Flower arrangements

Artificial flowers

It is true that fresh flowers make for gracious living but they do not last long in smoky, centrally heated and overcrowded atmospheres, such as may exist in busy bars and lounges, and their constant replacement is an expensive item. For these reasons the use of artificial flowers has become more wide-spread; added to this, the quality and design have improved to such an extent that it is necessary in many cases to touch the flowers in order to find out whether they are real or not.

Firms under contract will supply the whole arrangement, vase, artificial flowers and foliage, suitable for a particular position. For example, a contract may be made for the provision of one arrangement for the entrance hall to be changed monthly, and the flowers to be in keeping with the décor and with the time of year. Thus there is a saving of time and labour for staff, an avoidance of spills and no space required for a flower room. It must be emphasized that the provision of artistically arranged artificial flowers is not cheap, although over the year, it may be a little cheaper than the use of fresh flowers.

Dried and silk flowers, particularly, have their place but certain authorities regard them as a fire hazard and ban their use in public places unless treated with a fire-retardant finish.

21

PLANNING TRENDS

Guests may stay in an establishment for convenience, for pleasure or from necessity; in each case they will require comfort, good food and service and the provision of these is dependent on good planning and organization throughout the house.

As has been suggested earlier, management relies in many ways on the housekeeping staff for the impression *guests* receive of the establishment and consequently for their contentment, recommendations and, in the case of hotels, their return in the future. The organization of the department has already been shown to be of great importance, for if it fails there may be repercussions right through the 'house'. However, in many instances, the impression made on the *guests* can be improved, and the work of the staff be made much easier, by careful initial planning of the building, furnishings and fittings.

Alterations and reconstructions in a building can be expensive and difficult operations and there are places where for one reason or another, they cannot be carried out to any great extent. In these cases the best has to be made of a bad job and it is unfortunate that sometimes the shortcomings of the establishment's construction and furnishings are given as excuses for the poor or indifferent service offered to the *guests*.

Inconvenience and annoyance to *guests* and staff can be caused by:

different floor levels necessitating steps;
insufficient or badly placed lifts;
too small service rooms;
badly placed linen chutes or rubbish chutes;
unsightly plumbing or electric wiring;
insufficient lighting and electric sockets;
badly placed electric sockets;
inconvenient and badly placed furniture and fittings;
unsuitable surfaces for the wear and tear to which they are subjected;
unnecessary dust-traps;
insufficient washing and bathing accommodation.

In many instances, with a little forethought many of these disadvantages could have been overcome and it is now realized that initial planning, whether for a new or reconstructed building, is of tremendous importance and can benefit both the *guests* and the staff. It is for this reason that 'mock ups' and 'trial runs' are being used much more frequently now than formerly. A building should grow from its requirements, so every piece of plant, furniture or fitting should be planned or designed in a functional manner in order that it may accommodate the space available, withstand the abuse it may receive, be comfortable and convenient to the user (*guest* or staff) and be practical in cost.

In order to achieve centralization of services, for example lifts, emergency exits and chutes for linen and rubbish, due consideration should be given to the shape of a new building. This will obviously be influenced by the space available but some of the shapes used can be seen in Fig. 21.1 below; in these cases the positioning of lifts, fire escapes and chutes is particularly easy.

Figure 21.1 Different building shapes: Y,T, hollow round, hollow square and rectangular

It is usual for rooms to be on either side of a corridor, so that the maximum number can be provided in a given building and the window in each room will be in the one outside wall opposite the door. In hotels it may be an advantage to have communicating doors between one room and the next; it is then possible to connect several rooms if required, although it may complicate matters when considering the placing of furniture.

The recognized minimum area for a single room is approximately 9 m², a double room 11 m² and for a twin bedded room 13 m²; larger rooms will, possibly, mean extra plumbing, lighting, carpeting and furniture. The type of establishment will influence the size of the rooms to some extent and the number of rooms a maid would be expected to service. In a luxury hotel there will be a number of suites; the rooms may be larger than usual; the guests may require more personal service and the furniture and furnishings may need more attention. A maid, therefore, may only service 8 to 10 rooms whereas in a less expensive hotel, where the tendency is for smaller rooms and more easily cleaned furniture, a maid may be expected to service 12 to 15 rooms; and in other establishments where little personal service is given, maids may clean many more rooms. Facilities such as service rooms, linen chutes, rubbish chutes and ducts for the central suction cleaning system, should be given careful consideration so that the maids need to walk as short a distance as possible.

Lifts in many new buildings are fully automatic and travel at a great speed, and there are people who are very nervous of the speed, the confined space and using the lift without an attendant; some have even refused to go to certain hotels owing to the lack of lift attendants.

The *corridors* should be wide enough to enable the use of wheelchairs and trolleys, for people to pass comfortably and to prevent any feeling of claustrophobia. Steps can prove a great inconvenience for the use of wheelchairs and trolleys and where possible they should be replaced by ramps. Many corridors have little or no external light, and in order to prevent accidents adequate artificial lighting is necessary throughout the 24 hours; *guests* can then see their way clearly and the room numbers easily. In some buildings, during the day, borrowed light is provided on the corridors by having fanlights over the doors but, when this is the case, *guests* can be disturbed at night by the corridor lights. To conform with local by-laws, secondary lighting must be available in hotels to show up emergency exits. With modern methods of construction there are not the same fire risks in a new building, but under the Fire Precautions Act 1971 fire doors or fire breaks are necessary to confine a fire to one part of the building and to exclude draughts which might help spread the fire. There must also be adequate means of raising the alarm and of escape in case of fire (see pages 62–3).

Room doors are usually 76–90 cm wide and they and the architraves tend to be simple in design, so giving fewer ledges on which dust may settle; the simple design has the disadvantage of not breaking up a long corridor in the way that more ornate doors and architraves would. This can be overcome to some extent by the use of coloured doors. The outside of the door has the room number on it and in large establishments the number also indicates the floor, eg 101 is the first room on the first floor, 201 is the first room on the second floor, and in some hotels the number is also on the inside of the door for the convenience of the guests.

Door locks are of various kinds. They are expensive and it is important that they give effective security. Guests' rooms in most hotels have mortice locks which fit into a slot in the door frame and are concealed from view. It is advisable that this type of lock is spring operated so that the door locks when closed and a key is required to open the door from outside the room; the handle only is needed to open the door from inside. There may be a 'double throw' action when the bolt is sent further into the frame of the door by a second turn of the key. In a master key system this is done by the grand master key which locks the door against all other keys (room, submaster or master key), as well as giving extra security against the lock being prised open.

Guests may wish to have privacy and added security in their rooms and the lock may include a catch which can be operated from inside the room to prevent the door being opened by the submaster or master keys. The grandmaster does, however, override this catch and opens the door, a precaution necessary in case of emergency.

More sophisticated, more expensive and more secure lock systems are now being used for guests' rooms in some hotels, eg rekeying and computerized keyless lock systems (see page 57).

Peepholes (spyholes) which allow guests to see who is outside their door are now considered standard security equipment in some establishments. Safety chains may be considered a hazard in case of emergency.

To prevent the surface of the door becoming rubbed by the room key tab, finger plates or metal shields are often placed below the keyholes. Further marking can be prevented by *housekeepers* and maids knocking with their key on the door handle or metal plate instead of on the actual surface of the door.

Rooms generally have much lower *ceilings* than previously. During modernization, it is possible for ceilings to be 'dropped' by making use of plaster boards, acoustic tiles or other materials, thus improving the dimensions of the room, heat and sound insulation. With the lower ceilings there is no need to break the height of the wall with picture rails and pictures are hung with hidden cords. Many rails in the past held no pictures and were just another ledge which required dusting.

Walls should be as sound-proof as possible to exclude noise from the corridor and the next bedroom. Skirting boards, while essential to prevent damage to the wall, should not present a ledge which needs dusting; they may be slightly recessed, or coved when they may be made of thin metal strips, eg anodized aluminium or brass. (See Fig. 21.2.) The coved metal strip enables the edge of the carpet to be cleaned more satisfactorily by an upright vacuum cleaner. A doorstop is necessary in most rooms to prevent the walls from becoming marked.

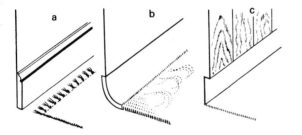

Figure 21.2   Different skirtings: a) conventional, b) coved metal strip, c) recessed

Modern *curtain tracks* have a better appearance and they and the runners are now usually made of rustless material eg plastic. They do not rust (and stain the curtains) or stick as the older metal ones were inclined to do, unless they were well maintained; in addition the plastic ones make considerably less noise when the curtains are drawn. With the improved appearance of curtain headings there is not the same need to have pelmets, although they may be a feature of interior design.

Windows should as far as possible be a standard size as this avoids the need for many spare sets of curtains and sorting curtains of different lengths. The ease with which windows can be cleaned should be given due consideration, and it is an advantage if both sides of the window can be dealt with from inside. Windows may be double glazed to provide either heat or sound insulation; for heat insulation the distance between the two panes of glass should be 0.6 cm while to counteract noise there should be a space of at least 11 cm, some sound absorbent material at the base between the panes and one pane should be slightly tilted so that vibration does not get across the gap. It is possible in this way for sound to be reduced by 85% and double glazing in establishments is normally used for sound insulation.

Fireplaces are becoming obsolete and as a result a useful aid to ventilation has been removed, but along with it has gone a source of dust and an entry for birds bringing soot and rubble with them. Where they remain, a coal fire can give a welcoming appearance in foyers, lounges and other public areas. Most new hotels have an air conditioning plant installed which enables the guest to regulate the filtered, warmed or cooled, and dried or moistened air entering the room and in cases where the windows should not be opened, notices to this effect should be clearly displayed (see page 274).

In other places, heating is normally by a central heating system which necessitates the use of radiators in the rooms and natural ventilation. The radiators are generally placed under the windows and enclosed behind a grille but the position and type of radiators or grilles need careful consideration from the point of view of appearance, ease of cleaning and efficiency of the sytem. Still other places rely on gas or electric fires.

Noise is reduced when floors are covered with wall to wall carpeting; at the same time appearance is enhanced and cleaning made easier and in any room the planning of the interior and the arrangement of the furniture and fittings have to be considered from both the functional and decorative points of view.

Planning in hotels

Suites of rooms are to be found in luxury hotels, consisting of one or more bedrooms, a private bathroom and a private sitting room. It is an advantage if these rooms can be let separately when required and with communicating doors more rooms may be added and the suite enlarged. In some luxury hotels there are also penthouses, which are still more luxurious than the suites and in four and five-star hotels each bedroom has a private bathroom en suite.

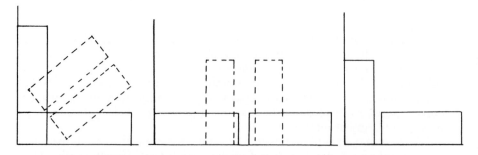

Figure 21.3   Different positioning of twin studio beds (dotted position at night)

Figure 21.4 Studio bed by day and night in single room

In some hotels there are studio rooms where the beds may be used during the day as settees and which are arranged to give maximum space so that the room may be used for leisure or business purposes (see Figs. 21.3 and 21.4 above). But in many hotels the majority of rooms are twin bedded (see Fig. 21.5 overleaf).

There are many hotels with few or no private bathrooms and in such cases there should be a washbasin in each room and an adequate number of conveniently sited baths and WCs for the number of guests. According to BS CP305 there should be:

WC: one per 9 persons, excluding occupants of rooms where WCs en suite;

Figure 21.5   Twin-bedded room

Baths: one per 9 persons, excluding occupants of rooms where baths en
          suite;
WHB: one per bedroom and one per bathroom and one in the vicinity of
          each range of WCs.

It is of course essential that at all times there is an adequate supply of hot
water.

Guests' rooms

It has been customary in many hotels to have more or less standard rooms
throughout, which it was hoped would satisfy the requirements of the guests.
The trend now, especially in three and four-star hotels is to up-grade and
provide a choice of more expensive rooms for the different types of traveller.
In addition to the usual presentation toiletries, sachets of detergent, mending
kits etc, these more expensive rooms may contain:

    remote control TV,
    video facilities,
    electric trouser press, which also takes skirts,
    hair drier,
    bath robes,
    fold-down drying rack,
    personal safe,
    tea and coffee making facilities (hospitality tray),
    mini bar,
    late night platters,
    early riser breakfasts,
    bottled water.

Some private bathrooms attached to these rooms will also have whirlpool
baths.
    A number of bedrooms may be designated for non-smokers as however

thoroughly a room is cleaned the smell of smoke permeates the carpet, curtains etc, and lingers in the room. Many non-smokers find this objectionable.

Multi-purpose rooms (originally studio rooms) are provided for business travellers. The rooms may then be used for leisure or business purposes and in the future may include computer terminals linked to the TV in the room.

Women travelling alone do not wish to be treated differently from men but they will appreciate rooms situated near a lift, that are designed with some thought for a woman's needs when travelling, eg good light for working or make-up etc.

An increasing number of hotels now equip some rooms with adaptations for the use of guests with disabilities, eg hoists, grab rails, entry phone, emergency devices and adequate space for wheelchairs.

Each type of room must be saleable to any guest so not all hotels go to the extent of providing all types.

Bedrooms

The appearance of the bedroom is affected by the size and relative proportions of the various pieces of furniture and fittings, and perhaps the article which has the most effect is the bed. As property prices increase and space is at a premium, there is a tendency for hotel bedrooms to be smaller and with lower ceilings than in the past, so frequently divans are used rather than bedsteads because they appear to take up less space. During the cleaning of the room, beds will require moving and the provision of castors or skids makes this easier as well as preventing damage to the carpet. Headboards are frequently fixed to the wall and one long one may be preferable to two separate ones in a twin bedded room, when there is the possibility of the two beds being put together as a double or moved to accommodate an extra bed.

It is convenient if a control panel for light, radio and TV switches is provided within easy reach of the bed. A vertical telephone may be fixed to the panel so saving space on the bedside table. On any type of telephone a small light operated by the telephonist may indicate a message is waiting at reception. Guests may contact the housekeeper, valet or floor service directly by telephone in many hotels. Attached to the headboard there may be shelves for use as a bedside table and for the storage of telephone books. However if the table is free standing there is the advantage that the beds can be converted into a double bed or rearranged to make room for an extra bed or cot. The bedside table should not be too high but of sufficient size to hold the early morning tea tray comfortably as well as other necessities for the guest. Breakfast trays can present problems to the guests; either they are lodged precariously on the bedside table or placed on the bed when their removal requires a feat of agility on the part of the guests wishing to get out of bed. Some hotels help the comfort of the guests by providing coffee tables, the top of which can be raised to go over the bed and hold the breakfast tray and when the meal is finished the table can be pushed to one side.

Dressing tables are frequently plain, flat surfaces which can be used as writing tables and so there should be sufficient knee space. Rounded edges of the high pressure laminates are an improvement over the traditional square edge in which the edging tends to peel off, and there is less resistance to knocks and less danger of injury to guests and staff. Rounded insides to the drawers make for easier cleaning and, if the bottoms of the drawers are made of laminated plastic, 'perspex' or some similar material, they need not be lined with paper, so saving time. The mirror may form part of the dressing table or a large one may be fixed to the wall above it.

One unit, a *built-in wardrobe*, with hanging space on one side, and shelves or open trays on the other, has been in use for some time but there is a move, particularly in three and four-star hotels, towards more individual pieces of furniture, when the advantages of shelves and open trays instead of drawers are lost. With shelves and open trays guests are less likely to leave articles behind; they can be more quickly and efficiently cleaned by the maids and checked more easily by the housekeeper. The wardrobe should be both long and wide enough, the coathanger rail not too thick and with sufficient space above to enable the hangers to be placed on it easily. Sliding doors save space, but they need to be well maintained and are not as dustproof as hinged ones. Any handles, locks and hinges should be strong enough to withstand the wear and tear to which they may be subjected. In motels and other places where there are many 'one nighters' it is sometimes thought unnecessary to have doors on the wardrobes.

In all rooms there should be a *full length mirror* which may be fixed to some convenient place on the wall or even fixed to the inside of the door so that guests may use it to its full advantage, particularly as they are leaving the room.

Luggage racks may be fixed or movable and should be big enough to take a

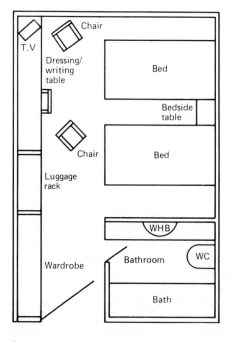

Figure 21.6   Twin-bedded room with private bathroom

large case comfortably; fixed ones may be cantilevered when they must be strong enough to withstand the weight of heavy cases or someone sitting on them. The surface of wooden luggage racks becomes scratched unless metal strips or a covering of ribbed rubber or similar material is put on them; horizontal stainless steel tubes form a suitable top which is not damaged by the bumping of cases nor will it damage the cases. Where possible some protection should be given to the wall or piece of furniture (often the foot of the bed) against which the luggage rack rests. Collapsible luggage racks are useful so that when guests are staying for some time and have unpacked, the racks and cases may be removed. A strong hook from which a portable wardrobe may be hung is an advantage to the guest who uses this type of luggage and prevents this being hung from unsuitable places. If luggage racks were made with a shelf underneath, the difficulty of knowing where to put the bedspread when removed from the bed at night would not arise.

There should be an upright *chair* or stool for the dressing or writing table and a more comfortable chair for each guest. Fully upholstered chairs become readily soiled and, if without castors, many are too heavy to move easily, consequently chairs with durable frames, webbing and fitted cushioned seats with removable covers, are more practical in bedrooms. Where there is sufficient space a small coffee table may be provided.

Television is a feature in most hotel bedrooms and the set is normally placed so that it can be seen from the bed and comfortable chair. (The socket should not be behind the set, dressing table etc, to deter the guests from switching off at the socket.) A remote control set is advisable and radio, TV and in-house movies can be available over a single cable. Hotel services can also be shown on the TV screen.

Eventually the TV set may double up as a VDU (visual display unit) with a keyboard catering for:

electronic shopping,
settlement of guests' accounts from the room,
guest information services etc.

Pay-TV systems are accepted all over the world and are aimed at the business guest who is likely to demand the very best. The systems available are pay per view or pay per day (generally noon to noon).

Tea and coffee making facilities and minibars are provided in many rooms and in some cases a disposable box or small refrigerator houses the necessities for a continental breakfast. The sales from the minibars may be automatically recorded in reception as is the case when telephone calls are made on the direct dialling system or the bars may be checked each day. Iced water may be available in some hotel rooms, while in other hotels there are ice-making machines on the corridors. Dumb valets, ie electric trouser presses which also take skirts, and shoe-cleaning machines are provided in some hotels.

Fire detectors and alarms and intercom-cum-baby-listeners may also be found in some hotel rooms.

Lighting in guests' rooms should be adequate but not too bright, and it is usual for there to be several wall, table or standard lamp fittings in a room, rather than centre lights. The lights should be controlled at both the door and the bedhead in order to prevent a guest from having to enter the room in darkness or having to get out of bed to turn lights on or off, for example. Bedside lights should be carefully positioned or be manoeuvrable so that the book is lit and not the top of the guest's head; dressing table and wash basin lights should illumine the guest's face adequately.

Such articles as waste-paper baskets, ashtrays and pictures, as well as any articles provided for the use of the guests, eg writing paper, and folders containing hotel and other information etc, will also be found in bedrooms or sitting-rooms.

Bathrooms

The increased tendency to provide *private bathrooms* (see Fig. 21.7) has led to a greater need for careful planning as the bathrooms are in many instances smaller than formerly (minimum size for a bathroom is approximately 185 cm square with a 168 cm bath, a wash basin and WC). To cut space still further in some cases there may be a shower in place of a bath.

Figure 21.7   Private bathroom showing one type of vanitory unit and WC

Bathrooms are frequently internal and arranged in pairs so that there may be a common duct for drains, water pipes and ventilating shafts accessible from the corridor (see Fig. 21.8).

When bathrooms are internal there must be extract ventilation but building by-laws do not stipulate the introduction of fresh air. Ventilation in a bathroom should always be given very careful consideration as the naked body is very susceptible to draughts and the ventilation shafts can also act as sound carriers. Electric light switches should be either outside the bathroom or of the cord type. Public bathrooms should have cord type switches; as many public bathrooms are not internal they rely on natural ventilation.

Bathroom *floors* should be hygienic, unharmed by water and of an easily cleaned material. In some cases a sluice hole is incorporated, which is a great

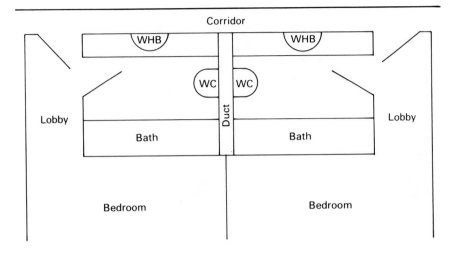

Figure 21.8 Two bathrooms side by side

help when guests let the baths overflow. The most satisfactory materials may be marble and ceramic tiles but they are very expensive and so seamless resin, vinyl and linoleum floorings are much more frequently used.

Baths are obtainable in various colours with soap dishes, grip handles and anti-slip devices sometimes incorporated. Baths are frequently 168–185 cm long by 70 cm wide, but vary in depth; there is generally about 30 cm of water to the overflow and the latter is approximately 14 cm below the top edge of the bath. There should be ample toe space at the bottom of the side panel to prevent scratching and make cleaning the bath easier.

Most private bathrooms have showers sited over the bath and these should be thermostatically controlled in order to prevent scalds. The height of the shower, if not movable, should be carefully considered. A soap container on the wall and a shower screen or a curtain of a suitable length to hang well inside the bath should be provided.

Wash basins match the bath in colour. They should be sufficiently large, the most usual size being 56 × 40 cm, but there are larger ones of 63 × 45 cm; there should be sensible soap-wells and sufficient room round and under the taps for ease of cleaning (see p. 23). There should be a mirror above the basin, a razor socket nearby and preferably fluorescent lighting so that the guest's face is well lit.

A shelf above the basin is not advisable as it is seldom large enough and there is the danger of articles dropping into the basin and cracking it, but adequate space for toothglass and personal toilet requisites is necessary. A vanitory unit consisting of a flat laminated plastic surface surrounding the wash basin satisfies the need extremely well (see Fig. 21.9 overleaf) and has the added advantage that it can be built in or cantilevered. Vanitory units may be found in bedrooms as well as bathrooms.

The *WC pan* is about 35–40 cm high and 60 cm deep including the cistern (wwp) which is about 50 cm wide; the WC pan may be cantilevered and

Figure 21.9   Vanitory unit

should be of the fully siphonic type which is not as noisy as the washdown one, is more effective and seldom goes wrong. The cistern may be in the piping duct and accessible from the corridor.

There is not often room in bathrooms for a stool, so the lid of the WC may be used as a seat. The sealing of the WC with a paper band has a psychological effect on the guests who consider such practices particularly hygienic. Toilet paper holders should be provided, and it is usual for there to be the two kinds of paper.

In many countries, *bidets* are the recognized fourth piece in a bathroom suite and they may be found in private bathrooms in the UK (see Fig. 21.10 below). They are about 38 cm high, 36–38 cm wide and project about 58–60 cm from the wall. Their main purpose is for the thorough washing of the anus and genitals, but some people find them useful as footbaths.

Other amenities which should be provided in private bathrooms for the guests include towel rails, which should be sufficiently far from the wall to

Figure 21.10   Bidet

allow ample space for the thickness of the towels (when the rails are heated they may be the sole means of heating the bathroom). Drip dry rails over the bath, a hook and a lock on the door should also be provided. Toothglasses may be disposable or when washed may be put into paper bags.

A telephone in the bathroom can save much irritation and annoyance to guests and its use may prevent the bath from overflowing (see pages 158–60 for further details of sanitary fittings).

Suites

In a suite the main essentials of the bedroom and bathroom are similar to those of any bedroom or bathroom in the hotel but usually the rooms are larger and have more luxurious furniture, fittings and furnishings. A bidet will almost always be found in the bathroom and often a whirlpool bath also.

The sitting room will contain a sideboard or cocktail cabinet from which drinks are served and meals may be provided by room service. Relaxed comfort is the predominant feature in these rooms which are normally close carpeted with tasteful furnishings and comfortable furniture. The lighting fitments are decorative and generally include table and standard lamps.

Lounges

Lounges are provided for guests who wish to spend time in places other than in their bedrooms and where they may be served with drinks, tea or coffee in a relaxed atmosphere. In city and transient hotels the lounge may be an extension of the foyer whereas in a resort hotel, in addition to the usual lounge, there may be particular rooms set aside for television, reading, writing and for games.

In a lounge the furnishings will be comfortable and restful and the chairs arranged for guests to be able to converse in small groups. The lighting should give an inviting appearance and chandeliers or lampshades may be an attractive feature. (See Fig. 19.7 page 253.)

Cloakrooms

A ladies' cloakroom (powder room) needs to be provided for the use of guests, many of whom will be non-resident, and their impression of this room may influence their judgment of the whole hotel. It is usual for there to be several individual WCs, wash basins in the form of vanitory units, large mirrors with good lighting, coat hanging space and a chair. Where individual towels of either linen or paper are provided there should be a receptacle for the soiled towels and there may be such supplies as tissues, aspirins, sanitary supplies etc available. In a first-class hotel the décor of the powder room may be luxurious, eg fitted carpet, expensive wall coverings and flattering lighting effect and in such a powder room there will normally be a cloakroom attendant.

Where large functions take place in the hotel, eg banquets, balls, conferences etc, special arrangements are made for the safekeeping of male and female coats, quite apart from the men's cloakroom or the ladies' powder room.

Conference rooms

Conference facilities are available in some hotels and these should include the latest technology with audio and visual equipment etc, suitable chairs and tables. Good ventilation and acoustics are very important.

Syndicate rooms are frequently required and these and the conference rooms should be multi-purpose so that they may be used for commercial and social functions.

Leisure facilities

These are already provided in some hotels and the demand for these health and fitness facilities is likely to grow. They are an added incentive for conference organizers to use the establishment and can generate trade from the local community as many hotels offering the facilities operate clubs with a membership fee. Sports and leisure facilities may include:

swimming pool,
gymnasia,
solarium,
sauna and jacuzzi,
games rooms (and games rooms for children),
recreational areas,
lounge area with bar,
changing rooms.

Massage, beauty treatment and hairdressing are also available in some places.

Motels

Motels are specialized hotels planned for motorists and geared to their requirements. They are therefore situated on or near a main trunk road, and they provide a central catering building and often chalet-type accommodation, with car parking facilities nearby (see Fig. 21.11).

The central building contains the reception office, bar, restaurant and maybe a coffee shop, while the guests' accommodation consists of a bedroom with a private bathroom. In most cases guests drive up to the reception office, pay in advance for the night and receive a key. They then drive to their room and park their car nearby and let themselves in, carrying their own luggage and are then free to leave as early as they wish. These advantages to the guests make security within the motel very difficult as no one checks their leaving or what they may take with them.

Motel rooms are normally provided with 'do it yourself' equipment, eg extra blankets stored in drawers, tea and coffee making facilities, and there may be shoe cleaning, vending and ice-making machines in the corridors in order to keep staff to a minimum; the keynote throughout the motel operation is 'convenience' for the guest.

Motor hotels and Post Houses are similar to motels in that they provide the service and facilities thought to be required by the motorist, but the accom-

Figure 21.11   Motel

modation is not of the chalet type nor is it separate from the central reception and dining block.

Apartotels

Also known as time-share properties, apartotels have become popular on the Continent and they are increasing in number in the UK. They offer basically the same facilities as in an hotel but the accommodation units are sold or leased to individual purchasers or time-share groups. The owners or lessees have a right to use their own accommodation for an agreed and limited period each year and for the balance of time it is available for letting.

Planning in hostels, university halls of residence and hospitals

In modern residential buildings, for various types of hostels and halls of residence, the canteen or dining room may be on the premises or it may be some distance away and this will affect the planning.

Usually each floor is similar and is a complete unit for a given number of bedrooms. When there is no dining room on the premises then a utility or communal kitchen/dining area (amenity area) may be provided for a certain number of rooms. This area may be 'open plan' with cooker, refrigerator, sink and draining board, as well as a table or tables, and some chairs. If there is a dining room on the premises the utility room usually provides a water boiler or a point for an electric kettle, a work top and maybe a cooker, and an

Figure 21.12   Utility room with laundry facilities

ironing board with a fixed electric iron on which there may be a time switch (see Fig. 21.12 above).

In either case there are frequently laundry facilities on the ground floor, or in the basement, complete with washing machines, driers and irons for the use of all the residents in the building.

In most places the residents' rooms will be single (though there are places with doubles) and may include wash basins which, if enclosed in a cupboard, should have adequate ventilation because of the damp produced from towels, face flannels and possibly washed articles hanging round the basin. On each floor there will be a wash room area which will include bathroom and/or showers, WCs, and wash basins.

For example, in one hall of residence there are on a floor 22 study bedrooms each with its own wash basin and the wash room area provides two bathrooms, two showers, three WCs and two wash basins.

In another, where there are no wash basins in the study bedrooms, the wash room area provides two bathrooms (one with WC and wash basin provided for the use of visitors of the opposite sex), three showers, three WCs and six wash basins for a floor of approximately 20 students.

The University Grants Committee recommend one WC and one bath for six students or one shower for 12 and one wash basin for 3.

The Department of Health recommends for junior hospital staff one bath or shower and one WC for 4–6 people (wash basins in rooms).

The Department of the Environment recommend for the elderly one WC for 2 persons and one bath or shower for 4 persons.

The wash room areas should be as conveniently placed as possible for the use of all the residents on a floor. The floorings in these areas are easiest to keep clean and hygienic if they are of an impervious material such as quarry tiles or terrazzo with coved edges. Showers are more economical than baths for water (a shower takes 12½–25 litres and a bath approximately 50 litres) and the heat of the water should be thermostatically controlled. Showers are popular and whereas in hotels they are normally fitted over the bath to save space, in other establishments fewer baths are provided and the showers are sited separately. Opaque curtains of a suitable material are necessary and to prevent accidents there should be a non-slip base to the shower. Soap-wells should be provided for both baths and showers; they should be well positioned, allow the water to drain away and be easy to clean.

In hostels and halls of residence for women it is essential to have incinerators or other hygienic disposal facilities in the wash room areas for the disposal of soiled sanitary towels. In the first case the towels are burnt and in the second they are treated with a germicidal fluid and the container, which is on loan, is changed regularly in accordance with the requirements of the establishment.

Study bedrooms

In most modern residential hostels or 'homes', the *rooms* are bed-sitting rooms or study bedrooms (see Fig. 21.13 below); the latter should incorporate a well-lit work table or desk with drawers, an upright chair and shelves for books. See also Fig. 21.14 overleaf.)

Figure 21.13   Typical hostel room

The *bed* may be 85 cm wide (100 cm is probably too wide for these rooms) and at least 190 cm long and the mattress may be 'dropped' into the base; this is harder wearing and has a tidier appearance in the room than when the conventional mattress and base are used. Sometimes drawers are fitted in the base for greater storage space. Residents in hostels frequently consider their rooms as their homes and they tend to arrange the furniture to their liking, so with this in mind it may be better to provide beds with headboards rather

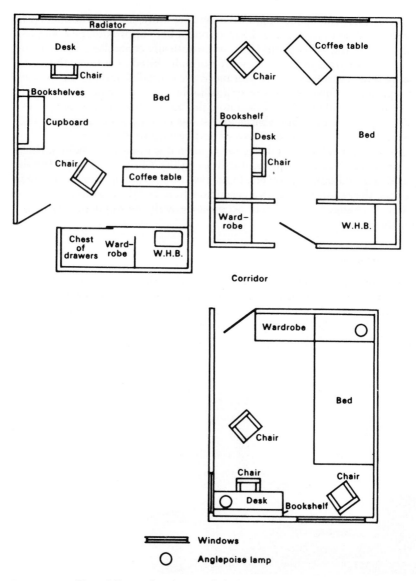

Figure 21.14   Three different plans for a study bedroom

than to fix headboards to the wall and the dressing/writing table is better to be free standing.

The *wardrobe* should be large enough to allow for bulky winter clothes and it should have a lock and key (the usual cupboard above may well prove useful for more permanent storage).

The *easy chair* should be durable, light weight and easily maintained but loose cushioned seats and backs of chairs are liable to be misused and get mislaid. A coffee table is often provided.

Soft furnishings are kept to a minimum for economy of maintenance; bed-spreads are normally of the throw-over type because they probably stand up to the harsh treatment that they are likely to receive better than fitted ones, and so ease maintenance.

The most usual *floorings* in the past were linoleum with bedside rugs, now vinyl flooring or fitted cord, tufted or adhesively bonded carpet may be used. Carpeting has the advantage of giving a warmer appearance, better heat and sound insulation and is easier to clean, especially if it has been given a stain resistant finish. *Wall surfaces* are frequently painted.

To avoid nails and Sellotape being used haphazardly on the walls, a pin board is useful; there should also be a firmly fixed towel rail and a good, strong hook on the inside of the door. Because of the inconvenience caused through residents locking themselves out of their rooms, deadlocks of the mortice type rather than the self-locking type of lock on the doors may prove more satisfactory.

The size of a single room varies considerably and may be from approximately 8–10 sq m. The rooms tend to be longer than they are wide and where students share rooms it may be possible to provide split level working and sleeping areas, so that a student working late is less likely to disturb the other. There should be adequate lighting in the rooms and this will be helped by the provision of sufficient sockets for reading lamps.

Cork floorings for the *corridors* have the advantage of being quiet and deadening the sound of footsteps. There should be no steps in awkward places but, where they are found, ramps should be arranged for the ease of people in wheelchairs and also for trolleys. The service core, lift well and escape staircase should be as near centrally placed as possible (see Fig. 21.15 overleaf). Wall surfaces on the staircase and corridors may be painted with the multi-colour paints for hard wear and ease of maintenance but there are other suitable materials.

Common rooms

In addition to the residents' own rooms there will be common rooms and often recreational rooms. In furnishing these rooms consideration should be given to ease of cleaning and the wear and tear to which they may be subjected.

The provision of coin-operated telephones and vending machines of various types is usual in all establishments.

Hospitals

In modern hospitals most of the above points are applicable to the 'homes' and 'residences' where the rooms are of the bed-sitting room type. Vinyl wall coverings and vinyl floorings with carpeting make for ease of maintenance.

As far as patient care areas in the hospitals are concerned there will be tremendous variation in planning; this will depend on the size of the unit and its particular use, eg private rooms, wards of varying sizes, day rooms, foyers, waiting rooms, offices, etc. While most parts are kept simple in design, with the work of the medical and nursing staff, the comfort of the patient, the ease of cleaning and the prevention of cross infection kept in

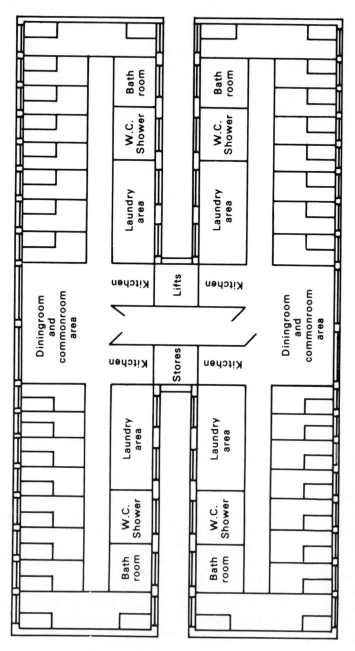

Figure 21.15 Floor plan of hostel with kitchen–dining area

mind, hospitals are now given a more inviting atmosphere than in the past by the use of warmer colours and textured surfaces, particularly in the public areas. Thus close carpeting, comfortable furniture and attractive soft furnishings are to be found in foyers, waiting rooms, staff coffee lounges and offices etc, in some of the modern hospitals. In other cases, PVC tiles and, especially at clinical levels, PVC sheet flooring with welded joints and coved skirtings may be chosen.

Where the built-in vacuumation system is in use there should be sufficient sockets in order to prevent excessive lengths of hose being used and stored.

Ceilings in the larger areas may be suspended to contain recessed lights and air diffusers, and acoustic plaster may be used on the walls.

In the larger wards, the bathrooms, toilets, sluice rooms etc, should be as centrally placed as possible to prevent patients and staff having to walk long distances. Elbow-operated taps are necessary in treatment areas.

GLOSSARY

Architraves   Mouldings round doors and windows.

Bedding   Term used for the articles on a bed, and normally includes the launderable linen.

Block tips   Share of tips, usually from tours, conferences etc, which the *housekeeper* distributes with discretion amongst the staff.

Cantilevered   Articles resting on a bracket projecting from a wall.

Ceramics   Articles made from clay, eg china and tiles.

Checkout   American term for a departure in an hotel.

Cleaning agents or materials   Include abrasives, detergents, solvents, polishes, etc.

Cleaning equipment   Includes brooms and brushes, electrical equipment, containers, cleaning cloths, etc.

Cutlery   Spoons, forks etc, as well as knives.

Departure   Room from which a guest is expected to leave or has already left.

Discards   Condemned articles in the linen room which may be renovated for other uses or used as rag.

Ergonomics   Study of people in relation to their working environment.

Floor seals   Semi-permanent finishes of cellulosic or plastic composition, applied to render a floor impermeable and to protect its surface.

Furnishings   Include soft furnishings, carpets and furniture.

General assistant   Employee who in a small hotel assists generally in any department.

In-house   On the premises, belonging to the establishment.

In situ   'On the spot' or 'on site'.

Linen   Material woven from flax, but the term 'linen' is often used loosely, to denote launderable articles found in the linen room.

OOO   Out of order.

Ready room   Room which has been serviced and is ready for re-letting.

Refurbish   To give a 'new look' to a room by redecorating, the renewing of soft furnishings and possibly the carpet, and the 'touching up' of furniture.

Re-sheeting   Putting out clean towels in a bedroom and making up the beds with clean sheets and slips.

Room state or occupancy list   List on which the assistant housekeeper states whether the room is vacant or occupied and, if possible, the number of sleepers in each room, and it is required by the receptionist and control office in a large hotel, at regular times each day.

Soft furnishings   Include curtains, cushions, loose covers, bedspreads and quilts, but not carpets.

Spread-over   Total number of hours over which a duty extends in any one day, eg 7 am–2 pm and 6–10 pm has a spread-over of 15 hours.

Textiles   Woven fabrics, eg cotton sheeting.

Turning down   Term applied to the work which maids may do each evening in guests' bedrooms in hotels.

Uniform   'Dress' of specified material, colour and design, usually provided by the establishment for certain staff.

Vacant room   One previously serviced and not yet occupied.

Vacated room   One from which the guest has left.

Water waste preventer (wwp)   WC cistern.

INDEX

INDEX